DATE			

Mechanical Metamorphosis

Recent Titles in
Contributions in American Studies
Series Editor: Robert H. Walker

A New World Jerusalem: The Swedenborgian Experience in Community Construction
Mary Ann Meyers

Musical Nationalism: American Composers' Search for Identity
Alan Howard Levy

The Dilemmas of Individualism: Status, Liberty, and American Constitutional Law
Michael J. Phillips

Sons of Liberty: The Masculine Mind in Nineteenth-Century America
David G. Pugh

American Tough: The Tough-Guy Tradition and American Character
Rupert Wilkinson

Uncle Sam at Home: Civilian Mobilization, Wartime Federalism, and the Council of National Defense, 1917-1919
William J. Breen

The Bang and the Whimper: Apocalypse and Entropy in American Literature
Zbigniew Lewicki

The Disreputable Profession: The Actor in Society
Mendel Kohansky

The Formative Essays of Justice Holmes: The Making of
an American Legal Philosophy
Frederic Rogers Kellogg

A "Capacity for Outrage": The Judicial Odyssey of J. Skelly Wright
Arthur Selwyn Miller

On Courts and Democracy: Selected Nonjudicial Writings of J. Skelly Wright
Arthur Selwyn Miller, editor

A Campaign of Ideas: The 1980 Anderson/Lucey Platform
Clifford W. Brown, Jr., and Robert J. Walker, compilers

Dreams and Visions: A Study of American Utopias, 1865-1917
Charles J. Rooney, Jr.

Mechanical Metamorphosis

Technological Change in Revolutionary America

Neil Longley York

FOREWORD BY
Brooke Hindle

Contributions in American Studies, Number 78

Greenwood Press
Westport, Connecticut
London, England

Library of Congress Cataloging in Publication Data

York, Neil Longley.
Mechanical metamorphosis.

(Contributions in American studies, ISSN 0084-9227 ;
no. 78)
Bibliography: p.
Includes index.
1. Technological innovations—United States—History.
2. United States—History—Revolution, 1775-1783.
I. Title. II. Series.
T173.4.Y67 1985 609.73 84-11845
ISBN 0-313-24475-8 (lib. bdg.)

Library of Congress Catalog Card Number: 84-11845
ISBN: 0-313-24475-8
ISSN: 0084-9227

First published in 1985

Greenwood Press
A division of Congressional Information Service, Inc.
88 Post Road West, Westport, Connecticut 06881

Printed in the United States of America

10 9 8 7 6 5 4 3 2 1

Copyright Acknowledgments

The author and publisher are grateful for permission to reprint the following materials.

Excerpts from the Papers of Tench Coxe in the Coxe Family Papers. Reprinted by permission of the Historical Society of Pennsylvania.

Letter from Samuel Osgood to John Adams of 23 October 1775. Reprinted by permission of the publishers from *Papers of John Adams*, Volume III, ed. by Robert J. Taylor et al., Cambridge, Mass.: The Belknap Press of Harvard University Press, Copyright © 1979 by the Massachusetts Historical Society.

To my parents
Eric Kingsmill York and Joel Barlow York
from a grateful son
and
to my grandfather
Irvin L. Barlow
a fellow lover of books

CONTENTS

Illustrations ix

Abbreviations xi

Foreword xiii

Acknowledgments xvii

Introduction 3

1. Home Manufactures and American Independence 8
2. The Isolated Inventor 37
3. War Economy: The Munitions Industry 63
4. Genius Frustrated: The Visionaries 87
5. Genius Recognized: The Engineers 112
6. Limits to Innovation: The Pennsylvania Rifle 132
7. Industrial Beginnings 155
8. The Inventor Celebrated 183
9. Inventing a Nation 213

Bibliographical Note 225

Index 231

ILLUSTRATIONS

1. Sawmill in colonial New York 12
2. Boston home manufactures notice 19
3. Arthur Donaldson's river dredge 50
4. Portrait of Christopher Colles 52
5. Christopher Colles's New York water supply proposal 54
6. Massachusetts munitions production notice 66
7. A committee of safety musket 71
8. Model of David Bushnell's "Turtle" 93
9. Schematic drawing of David Bushnell's submarine 94
10. Faden map of the Delaware River 119
11. A Pennsylvania rifle 134
12. Portrait of Daniel Morgan 144
13. Portrait of Tench Coxe 159
14. Portrait of Samuel Slater 169
15. Portrait of Oliver Evans 188
16. Oliver Evans's automated flour mill 191
17. John Fitch's steamboat 198

ABBREVIATIONS

References in the notes that are abbreviated after the first citation are listed here. Other references that are repeated in the notes appear in shortened form.

APS	American Philosophical Society
GWP	George Washington Papers
HSP	Historical Society of Pennsylvania
JCC	*Journals of the Continental Congress*
LC	Library of Congress
PCC	Papers of the Continental Congress
PMHB	*Pennsylvania Magazine of History and Biography*
PRO/CO	Public Record Office, Colonial Office
WMQ	*William and Mary Quarterly*

FOREWORD

The American Revolutionary Era was a seminal period in the development of technology and of American attitudes toward technology. The political, social, and even religious changes resulting from the Revolution have long been recognized and have received continuing attention. The nature of the newly designed government and the directions given to the newly emerging American civilization have not been overemphasized. This was the foundation period of the nation, and, as such, its history has permanent importance. This history, however, includes close to its center technology and the great changes relating to it that came out of the Revolution, developments not recognized much previously.

It is no longer possible to neglect these foundations of our present world. The character and quality of American life today owe at least as much to our celebration of technology and to the industrialization begun at that time as they do to the conventionally understood American Revolution. Colonial America made use of a large variety of technologies; competent craftsmen worked in their shops; and American farmers, isolated as they were, became adept in numerous skills. They did not, however, give clear evidence of special interest in or capability for carrying through significant technological change. Occasional inventions and innovations emerged, but, in this realm of achievement, the colonies lagged far behind the leading nations of Europe. The Revolution changed this. Those who carried the Revolution through did indeed build "a technologically oriented society."

As with the political revolution, the most fundamental change occurred in the hearts and minds of the people. In this case, only a few of the people were involved in the beginning, but some enthusiasts for technological change became active with the rise of Revolutionary fervor, before the War of Independence. Thinking in new directions, they grew concerned about the enormous extent of their undeveloped land and resources and by their lack of the physical means, the capital, and the labor required to develop it.

They had little capital and little ability to apply it; their sparse population was scattered widely with very deficient means of transportation and communication. They were colonial dependents not only politically but commercially and economically as well. But, there might be an answer if the Americans could introduce the new machines, the factory system, and the new labor and management patterns emerging in a movement later tagged the Industrial Revolution in the mother country.

Early and largely unsuccessful efforts to import this new technology were interrupted by the War of Independence, which, however, brought its own preemptory demands for technological development. The war effort exposed another large realm in which the Americans were undeveloped and unprepared. It similarly encouraged technological solutions to repair the deficiencies.

Most related to the attempt to import the machines and techniques of the Industrial Revolution were those efforts to make products demanded by the war. These were not a matter of innovation but of establishing production facilities already well developed in Europe and not previously needed in America. Most conspicuous was the production of saltpeter and the manufacture of gunpowder from it. Also established were gunlock factories, cannon foundries, and salt works. The social and political support for and the impediments to these endeavors turned out to be decisive. Overall, the results showed limited and temporary successes. In the critical matter of gunpowder, most of it in the end had to be imported.

Another category of need, that of military engineering, was raised to a high level by the effort to substitute technology for military deficiencies. After Bunker Hill, the Americans, conscious of their lack of tactical mobility and their incompetence in bayonet combat, emphasized heavy fortifications. As soon as they could, they imported continental engineers for these works, but one of their most dramatic complexes, the Delaware River defenses, was largely an American design. This story benefits, as do other episodes told here, from a close understanding of the many facets of the effort.

The attempt to apply technology to reduce the advantages of a militarily superior foe was pressed especially in weaponry. A multitude of proposals was put forward, and a few were tried, but the overall results were very limited. Most important were the Pennsylvania rifle, once called "the secret weapon of the Revolution," and David Bushnell's submarine. The submarine came close to success, and the rifle had a history of success and failure that has not previously been so well clarified.

At least some Americans emerged from the war with a heightened technological consciousness. More than ever, the development of the country and the fulfillment of its potential seemed a monumental task. The feeble means available could certainly be strengthened by applying the new technologies and techniques of the English Industrial Revolution. Moreover, that had become more feasible with the resumption of immigration from the mother country and the influx of mechanics familiar with the new machinery.

Aspects of this story have been told before, but most accounts retain the initial attitude toward the process. They interpret technological change as an eponymic process in which improvements resulted from individual inventions or creative acts. The interest has been largely limited to the hardware, until relatively recently. The importance of the technology itself is not to be minimized, but it cannot be viewed with blinders that block the complex social, economic, and cultural dimensions of the history. These interrelationships are fully comprehended in this book and constitute its great strength. The technology at the center has to be understood, but it must be understood within its larger context.

Neil Longley York begins with a clear picture of the technology, which he is especially successful at placing, along with the inventors and operatives, in its social context. It becomes clear, for example, that Samuel Slater's success in introducing mechanized cotton spinning to the United States involved much more than building the machines in Pawtucket. Those who carried through such successes hoped they could and believed they should enrich themselves by acts that would benefit the country and the rest of humanity. America became peculiarly receptive to technology and was not disadvantaged, as is sometimes suggested, by the predominantly rural nature of the country. Farmers had a broader experience with many technologies than did their European counterparts, and city artisans were often less skilled but also less tied to a single operation. Moreover, American skills continued to be enhanced by those who brought in the latest European advances.

Governmental actions sought to encourage the new technology, but the most important encouragement was the state of mind. This was evident by 1790, although it has not yet been recognized even by most of those who study the period specifically. The Americans had begun not only to adapt but to invent and innovate on their own. The effect was not yet extensive, but the course had been set toward increasing industrialization and increasing reliance on machines. The change that had taken place was fundamental and penetrated all aspects of life. Not limited to those who worked with the machines, it was a product of the Revolutionary period. Once this is perceived, our understanding of the American Revolution will be rounded out, and our understanding of the role of technology in later American history will be improved.

Brooke Hindle
Senior Historian
Smithsonian Institution

ACKNOWLEDGMENTS

I was most fortunate to have had help with this book. Carroll W. Pursell, Jr., of the University of California, Santa Barbara, first encouraged me to pursue the subject at hand and allowed me to draw from his wide-ranging expertise. David Ammerman of Florida State University gave an earlier version of the text a close reading, and David Hounshell of the University of Delaware offered helpful comments on one chapter. I have long benefited from my association with Frank W. Fox, who has guided me for over a decade—first as his student, now as his colleague. Brooke Hindle, Senior Historian of the Smithsonian Institution, gave unstinting support; his many suggestions were incisive, and needed. Silvio Bedini and Terry Sharrer, also of the Smithsonian, were kind enough to loan me materials they had gathered for their own work. Robert M. Speck generously sent me information on Bushnell's *Turtle.*

Funds from the University of California graduate council enabled me to do research at the American Philosophical Society library, the Historical Society of Pennsylvania, the Henry E. Huntington Library, and the campus libraries of the University of California at Berkeley and Los Angeles. The University of California, Davis, assisted serendipitously by inviting me to teach in its Washington, D.C., program during the summers of 1980 and 1981. When not in the classroom, I spent my time leafing through manuscripts in the Library of Congress.

I owe a debt of gratitude, poorly repaid here, to librarians and archivists who lent their assistance, especially Stephen Catlett of the American Philosophical Society. I am also grateful to the University of California and Brigham Young University students who have had my ideas tested on them. Marilyn Webb and the typists at the BYU faculty support center shepherded the manuscript for this book through numerous—perhaps to them seemingly innumerable—revisions. At Greenwood Press, production editor Louise Hatem and copy editor Sandra J. Wendel caught scores of technical errors that had escaped my notice.

My greatest debt is to my wife, Carole Mikita York, who endured my pouting and impatience and occasional flights of fancy. As loving confidante and audience of one, she made the rewriting and revising less tedious. Our daughter Jennifer was spared from any involvement in this enterprise.

Part of Chapter 3 appeared previously as "Clandestine Aid and the American Revolutionary War Effort: A Re-Examination," *Military Affairs* 43 (1979):26-30; and much of Chapter 6 was first published as "Pennsylvania Rifle: Revolutionary Weapon in a Conventional War?" *Pennsylvania Magazine of History and Biography* 103 (1979):302-324. Material from both is used here with the permission of the editors.

INTRODUCTION

Americans have been called "a mechanical people," a people who boast of their "inventiveness" and "ingenuity." Indeed, Americans for the better part of two centuries have been celebrating "Yankee ingenuity" as a unique national trait. American pride is very much tied to a belief in American ingenuity and a sense that Americans have used their machines to build a better society. All too often American inventors and American-made machines are offered up as symbols of national greatness, even as evidence of national superiority.

Machines and mechanization, inventors and inventiveness, are part and parcel of the history of technology. The history of technology as it is now pursued cannot be neatly categorized. It is not simply a form of economic history, though it necessarily deals with economics. Neither is it an appendage of the history of science, though modern American technology is very much tied to science-based industry. The history of technology cannot be easily defined because it is concerned as much with the reasons behind the substitution of machine power for human labor as it is with the details of particular technologies. While machines are the focal point of the history of technology, those machines are most significant when viewed within a social context.[1] Eugene Ferguson, Brooke Hindle, and Hugo Meier—to name just a few—have shown that the mechanization of American life has been a social movement in the broadest sense, a coming together of many individuals and institutions.

Tracing the mechanization of American life is therefore no simple task. Attempts to mechanize during the Revolutionary Era, 1760-1790, nicely illustrate the complexity and uneven rate of American technological growth. The Revolutionary Era was a time of tremendous change, not just in politics. There were technological changes as well, although they were not as obvious as the political changes. If one were to compare the American colonies in 1760 with the new American nation at 1790 in the hope of finding hordes of new machines and new methods of manufacturing, one would come away disappointed. The American colonies were part of a pre-industrial world; so, to a large extent, was the new American nation. There were no great manufacturing achievements, few great mechanical inventions, and certainly no sudden Industrial Revolution occurred. But there was a hint of things to come, a glimmer of fundamental change in the offing. The United States may have entered and left the Revolutionary Era as a pre-industrial nation, but it did not pass through unchanged. Even if no radically new technologies were introduced during the Revolutionary Era, some Americans became more technologically attuned. If by 1790, as Douglass C. North has noted, the stage had been set for far-reaching economic changes that came in the next century, those later economic changes were enhanced—and perhaps in part made possible—by the shifting technological outlook of Americans during the Revolutionary Era.[2]

I examine that shifting outlook in this book. Even so, I do not present an exhaustive, comprehensive examination of technology during the Revolutionary Era. Instead, I have selected topics that I feel accurately depict the social nature of technology and technological change and the American pairing of political and technological goals. While most studies of early American technology concentrate on the ingenuity of individuals, I am more concerned with the social response to that ingenuity. Most important, I have attempted to show how the response of American revolutionaries differed in given situations, depending on socioeconomic conditions and the technology itself.

What follows is very much a story of nationalistic aspiration. The technological movement championed by leading members of the Revolutionary generation was to some degree an outgrowth of, and dependent on, the move for political independence of the same years. Later, by doing its part to bolster the economy of the new nation, the movement for technological independence helped insure the success of the American political experiment. Naturally not all Americans were actively involved in either movement. But if the technological movement did not involve everyone, neither did it have many opponents. Ironically enough, some who rejected the political move for independence supported the movement for technological change, at least initially. Visions of technological change, then, fired the imagination of a variety of Americans, some of whom were politically at odds. Although those who were caught up in visions of future technological greatness were a minority, they were a significant minority because they charted the course for the new nation.[3]

They were not always able to follow the course that they had charted, however. They found that large-scale technological change was often difficult to bring about because success hinged on many factors: dependable sources of raw materials, a flourishing economy, available investment capital, demonstrated market demand, suitable machines, inventive craftsmen, imaginative entrepreneurs, a government willing to promote technological innovation, and a public enthusiastic about the prospects of technological change. The absence of one or another of those factors could retard technological growth. Moreover, the adequacy or inadequacy, appropriateness or inappropriateness of a technology depended on what was expected of it. Advocates of technological change often have broader social goals in mind, and leading Revolutionary Americans were no exception. They felt that some existing technologies were inadequate, inappropriate for their political objectives. They tried to create a more complex technological order that would further their nationalistic ambitions. The history of mechanization in Revolutionary America is in a sense a history of efforts to create that order. That new order, that higher level of technological consciousness, was necessary for the public identification with technology as a national life force, a mark of national distinctiveness.

The separate but mutually dependent factors determining technological change were not always easy for Americans of the Revolutionary Era to identify, particularly since they lived in an age when fine distinctions between pure science, applied science, and technology were rarely drawn. Scientific societies endeavored to encompass them all under the rubric of science and the "arts" or "mechanics." And, it should be added, the free exchange of scientific information on plants and animals that characterized the "natural history circle" of colonial scientists and their European contacts did not always carry over into the field of technology. Exchanging information on plants and animals did not as obviously endanger the imperial status quo. Technological exchange, unavoidably one-sided in many respects—to the benefit of the colonies and detriment of the mother country, was viewed as more dangerous to mercantilistic arrangements. Therefore, the more Americans attempted to mechanize, the more they threatened to disrupt the empire.

Because Americans poured much of their mechanical energy into manufactures, I have devoted considerable space to their factory experiments. Manufacturing required a more sophisticated approach to technological change than did isolated tinkering. Tinkering did not necessitate immense capital outlays and could be conducted informally by trial and error. Manufactures demanded much more. Technological advance through the former could be so unobtrusive, so perfectly utilitarian in answering a simple need, as to be almost unnoticed, allowing change to slip in through the back door. Technological changes within traditional handicraft trades were usually less shattering, less unsettling, than technological changes that altered the very nature of those trades. Manufacturing was more complicated and therefore more difficult. It forced Americans to take a harder look at their existing technological expertise as well as their ultimate technological objectives.

Fully half of the book deals with the War of Independence itself. The war tested and revealed a side of early American technology and public attitudes about technology in a way nothing else could. These chapters on munitions production, the efforts of inventors, engineering on the Delaware River, and use of the Pennsylvania rifle illuminate certain technological aspects of the war. Each, in a different manner, shows the possibilities of and limitations on technology in Revolutionary America.

The Revolutionary Era, after all, came simultaneously with the early stages of the Industrial Revolution in Great Britain. Before then, invention, technological innovation, and manufacturing had seldom been brought together in any systematic fashion. Only with the Industrial Revolution did manufacturing technology become the source of reverential awe; only then did inventors, the masters of mechanical improvement, become national heroes. Americans wanted to import certain elements of the Industrial Revolution, but they were not always able to capitalize quickly on the machines and techniques they borrowed from Britain. Their society before 1790 was simply not ready to foster large-scale technological growth. For many of the same reasons, invention was not seriously appreciated by more than a few Americans until 1790 and the close of the Revolutionary Era, when British machine technology began to rouse American industry. Inventors before 1790 operated on the periphery of the economy, and the public largely ignored their work. Most Americans had little use for inventors; they were not especially enamored with the "newfangled" and were often reluctant to adopt the latest devices.

By 1790 American attitudes had begun to shift. The more the American public came to crave technological change, the more inventors rose in their estimation. Inventors were more highly regarded by the public because they designed the machines that made technological change possible. Therefore in the chapters that follow on invention in the Revolutionary Era, I have examined the part played by invention in the metamorphosis of American technology. The changing attitude toward invention during the Revolutionary Era is a useful barometer for gauging the increased public interest in technological change detectable by 1790, when pre-industrial America took its first halting steps into the industrial world.

Though each chapter looks at a different aspect of technology, all are tied together by one major premise—namely, that we must not cloud our thinking about technology in early America by making anachronistic judgments or by basing our assumptions and conclusions on observations made in our own age of corporate capitalism. The "givenness" of modern technology (to borrow a term from Daniel Boorstin and use it in another context), the infusion of technological values into thought and behavior, did not come overnight with the founding of Jamestown in 1607. Likewise, Colonial Americans may have been part of a western culture that, as Lynn White and Lewis Mumford contend, had long ago given itself over to technology, but that surrender was only half-conscious. Americans in the Colonial Era tended to take technology for granted and did not attach to it the sort of social significance that later generations would. Our national love affair with machines came as the offspring of developments first prominent in the Revolutionary Era. The American Revolution, in short, was a quest for technological as well as political independence.[4]

NOTES

1. Lynn White, *Medieval Technology and Social Change* (New York: Oxford University Press, 1962), p. 28, noted: "The acceptance or rejection of an invention, or the extent to which its implications are realized if it is accepted, depends quite as much upon the condition of society, and upon the imagination of its leaders, as upon the nature of the technology itself." Also see S. C. Gilfillan, *The Sociology of Invention* (Cambridge: MIT Press, 1935), passim; Nathan Rosenberg, "Factors Affecting the Diffusion of Technology," reprinted in his *Perspectives on Technology* (Cambridge: Cambridge University Press, 1976), pp. 189-210; and Eugene S. Ferguson, "The American-ness of American Technology," *Technology and Culture* 20 (1979):3-24, for an insightful discussion of the many motivations for technological innovation.

2. Douglass C. North, *The Economic Growth of the United States, 1790-1860* (Englewood Cliffs: Prentice-Hall, 1961).

3. Carl Siracusa, *A Mechanical People* (Middletown: Wesleyan University Press, 1979), points out quite rightly that comments about "technological consciousness" have to be made carefully. Many people go along with technological change not because they are technologically "aware" but because it complements their preoccupation with success in the marketplace. Also see David Potter, *People of Plenty* (Chicago: University of Chicago Press, 1954), pp. 78-90, for Potter's discussion of "The Nature of American Abundance." Potter points out that change comes through the application of human skills to "raw materials," materials that become "natural resources" only after they are seen as such. Compare this to Friedrich List's observations in a letter to Charles Ingersoll of July 18, 1827, in Margaret Hirst, *Life of Friedrich List and Selections from His Writings* (1909; reprint ed., New York: Augustus M. Keller, 1965), pp. 192-193.

4. John F. Kasson's important book, *Civilizing the Machine* (New York: Grossman/Viking, 1976), p. 3, noted that the new nation "began not one revolution but two." Kasson devotes little space to the Revolutionary generation, however, except to discuss Thomas Jefferson's technological ambivalence and Tench Coxe's technological enthusiasm.

1 HOME MANUFACTURES AND AMERICAN INDEPENDENCE

By the second half of the eighteenth century the American colonies had matured visibly, to the gratification of most colonists and consternation of some British mercantilists. The colonies had long outgrown their status as mere "plantations," as European bridgeheads in the New World. Their growing sophistication in government and politics, cultural life, and economic pursuits threatened to disrupt imperial relations that had never been fully defined. Colonial American technology was caught up in the political disputes that followed. Those colonists frustrated with imperial relations were often, as a consequence, frustrated with the state of their technology.

Technological frustration became potentially dangerous to the empire as some colonists grew more politically ambitious. The further those colonists drifted from the mother country politically, the greater became their concern with their technological expertise. For they saw correctly that the weaker they were technologically, the weaker they would be politically; if they were not capable of sustaining themselves technologically, they discovered, they would not be able to do so politically either. Strained imperial relations caused them to look intently at their technological capabilities. Given the new role they expected their technology to play, it began to appear all the more inadequate.

Advocates of technological change wanted to improve, but they were not always able to because they could not work any sudden technological miracles. What is more, they were ready for a political break before they were ready for a technological break. They paradoxically followed Britain's technological lead, even as they attempted to build an economy that could stand on its own. The eighteenth century may have been "running mad after innovation," as Samuel Johnson put it, but—technologically, at least—American society understandably ran behind the mother country.[1] The colonists "mechanized" in some areas faster and more completely than the British—the proliferation of sawmills along the Atlantic Coast serves as a prime example—but in other fields they clung stubbornly to tradition or indulged in regressive rather than innovative practices. Those agitating for technological change therefore had a twofold task. First, they had to develop new economic enterprises, encourage the technological innovators among them, and continue importing new machines and production techniques from the mother country. Second, they had to persuade their countrymen to follow their lead. They had a difficult time with both.

Most colonial farmers, for instance, rarely made use of the latest tools. Neither did they follow the most progressive cultivating techniques that had become popular among the gentry in the mother country. It appears that no more than a handful of progressive farmers in America imitated the English gentry, joined this pastoral reawakening, and took their cue from imaginative farmers on both sides of the Atlantic.[2] Those "imaginative" farmers gave very clear directions. Jared Eliot of Connecticut called for more scientific farming in the colonies with his *Essays Upon Field-Husbandry*, which put advanced English practices into American terms.[3] Colonial magazines and newspapers repeated Eliot's challenge, reprinted extracts from British journals on proper agricultural techniques, gave news of local experiments as well as those in the mother country, and printed advice from prominent planters.[4] The works of Arthur Young, pre-eminent English agronomist, were available in American bookshops by the 1770s, as were the essays of other leading authorities.

George Washington was among the select few who attempted to practice the progressive, scientific techniques advocated by Young and the rest. By 1760 Washington had experimented with rye, clover, and peregrine species of wheat to escape his economic dependence on tobacco. A believer in soil conservation, Washington likewise experimented with various compost mixtures and river mud and assiduously practiced crop rotation to replenish exhausted soil. In a special botanical garden he grafted cherries, plums, peaches, pears, and grapes. This tidewater aristocrat broke the ground for his neighbors by adopting the newest and most advanced farm implements available and by designing some of his own. So did some New Englanders for their region who used the "cradle," a long-handled scythe, to triple the amount of grain they could mow. In addition, progressive farmers in some parts of Pennsylvania used new tools and techniques and diversified into livestock raising to supplement their cash crops.

American-grown grain did well commercially, as did indigo and tobacco, although the market for the latter became glutted by the 1760s. Yet if by the 1770s agricultural products were a leading export commodity, particularly in the Middle and Southern colonies, it was because of the fertility of the soil and availability of new land, not because of the expert or progressive practices of colonial farmers in Pennsylvania or anywhere else.[5] Washington and others like him who heeded the admonitions of Eliot and Young were a distinct minority. The average farmer did not have to be "scientific" to be successful.

If the typical farmer in Colonial America was not especially imaginative, he often took his grain to be ground into flour by a miller who might well have been very innovative. Water-driven grist and flour mills dotted the colonial landscape. Enterprising Virginia planter-merchant Roger Dixon advertised in 1770 that he owned a mill capable of processing twenty thousand bushels annually.[6] Dixon was one of many such entrepreneurs. Other Virginians managed smaller operations, sometimes combining their flour mills with a sawmill and cooper or blacksmithing shop. The Ellicott family of Maryland began a cooperative enterprise for wheat growing and milling outside Baltimore in the early 1770s, and their operation emerged as one of the most intensive and profitable along the Chesapeake. Pennsylvania had even more mills than Maryland and Virginia, and mill owners there noted smugly their location in "prime wheat country." Merchant mills with two water wheels, two or three pairs of "good French Burr stones," and adjacent sawmills were common in the outlying districts of Philadelphia, with Philadelphia County alone having 107 mills by 1773.[7] Local craftsmen made some of the hardware for those mills, such as a Chester County stonecutter who specialized in "mill stones of any dimension for grist mills" and a Philadelphia cutler who did ironwork.[8]

Millers who processed grain used dozens of tools. If most farm implements were rather small and simple, mill machinery could be quite large and complex. Some tools were imported, some were homemade copies, and others were assembled from original designs. Nevertheless, widespread use of tools did not mean that colonial millers always used the "latest" or that they were caught up in a conscious quest to develop ever more laborsaving devices. If some millers, like some sawyers, used new tools and new techniques to further mechanize their operations and increase profits, many others remained content to use "old" tools and "old" methods, just as many farmers used only the most basic tools and unimaginative planting and harvesting techniques. They could see no point to investing in improvements, particularly if there was no market pressure to do so.

Though commercial agriculture returned a good profit to farmer and miller alike, neither farmers nor millers in general seemed to have been concerned with advancing the rate of technological change. Farming remained primitive and soil intensive because of cheap land and more lucrative investments elsewhere. Farmers found that they could produce more than enough for subsistence and surpluses for market without taking steps to save the soil from exhaustion. Most only improved the land or their practices to satisfy immediate profit incentives. They rarely used the most advanced farm tools such as harrows and drill-plows made in England; they designed new devices of their own even more rarely. One hundred fifty years after Jamestown many Southern planters relied on axes, hoes, and shovels to clear, plant, cultivate, and harvest. Far from improving on updated English practices, most colonists imitated the "wasteful" yeomen rather than the "conscientious" gentry.[9]

South Carolina's lieutenant governor, William Bull, noted in 1773 that the resultant quick depletion of soils by farmers in his province made imperative new land grants from the Crown.[10] The message implicit in Bull's observation disturbed colonial critics of mercantilism. Patriots who were worried about the politics of empire also became worried about the state of their farming. If Bull was right, they would be in even worse shape when the Crown and colonial governments had no more land left to distribute. Improved farming techniques thus figured prominently on the list of "necessities" drawn up by those in the 1760s and early 1770s seeking to better colonial life. Colonial agitators who were worried about the future began to look for technological solutions to their economic problems and, eventually, to their political problems as well.

Unimaginative farming practices disturbed colonists who were discontented with imperial relations because poor planting and harvesting techniques were seen as symptoms of a malfunctioning empire. Concerned Americans believed that they had to upgrade their economy to protect their traditional freedom from imperial political intervention. Colonists distressed by what was happening on the farm were similarly concerned with the technological state of the trades. Advocates of change studied their handicraft industries and decided that they, too, had to be improved. While they did not become overly concerned about milling, they did worry about other trades; millers were usually more than able to hold their own, but the advocates of change feared that other tradesmen were not.

Without doubt some trades had fared well within the empire. By midcentury Lynn, Massachusetts, had gained fame as a shoemaking center, and Lynn cordwainers sold their wares in New York and Philadelphia as well as in Boston.[11] And Lynn cordwainers were not the only craftsmen besides millers and sawyers who prospered. Indeed, a thriving community of artisans had developed in the Middle and Northern colonies.[12] Nevertheless, most colonial craftsmen—sawyers with mechanized sawmills included—pursued trades that nicely complemented the mercantilistic blueprint of the mother country. Taken alone, they posed no threat to the empire. Other, larger-scale pursuits did, and it was to these that the advocates of change first turned. Moreover, technological agitators then wanted to tie smaller, simpler trades like shoemaking to larger operations like ironwork and present a united front to imperial administrators. In that context, all of the trades became a threat to mercantilism. The agitators pushed for home production, home consumption, and

1. Sawmill in colonial New York, symbol of early American ingenuity and productivity. Courtesy of the Library of Congress.

export—more shoes, more cloth, more iron. Their emphasis on increased productivity and greater competitiveness challenged the very nature of imperial relations.

By 1775 the iron industry in particular had therefore become a bone of contention, pushing some disgruntled Americans further away from the mother country. Even though the colonies produced a substantial amount of raw iron and did so with the blessings of mercantilists in London, colonial agitators demanded more. They hoped to expand existing facilities and move into new types of production; they wanted to make iron and steel tools, not just iron bars and pigs. Their desires ran at cross-purposes with the plans of most mercantilists back in London who wanted colonial ironmasters to restrict their activities. While imperial administrators wanted the colonial iron industry to prosper, they wanted it to do so within mercantilistic limits.

Pennsylvania had emerged by mid-century as the leader in commercial iron production, supplanting the earlier lead held by Maryland and Virginia. Pennsylvania, like New Jersey and parts of New York, had all the natural resources necessary for iron manufacture: surface outcroppings of ore, easily mined beds and bogs, plenty of wood for charcoal fuel, and limestone for ridding the ore of impurities.[13] Pennsylvania may have led the way, but New Jersey ironmasters were not content to be left behind. The Sterling Ironworks, overlapping Orange County, New York, and Bergen County, New Jersey, produced five hundred tons of pig iron annually, an amount rivaling that of the larger Pennsylvania operations.[14] American-made iron from New Jersey as well as Pennsylvania and Virginia found its way to Britain, where it received favorable attention. By the outbreak of the War of Independence, colonial iron operations reportedly produced more pig and bar iron than all of the forges, furnaces, and bloomeries in Wales and England combined. It has even been claimed that they accounted for as much as one-seventh of the world's supply.[15]

Some of the forges and furnaces that produced this iron were ambitious operations. Ironmaster Peter Hasenclever, in fact, made one of the few attempts "at what might be called large scale industry in the colonies." He purchased over fifty thousand acres for his "American Iron Company" in New Jersey and New York. From 1765 to 1767 he built five blast furnaces and seven forges, worked by foundrymen and artisans brought over from Germany and England.[16] Maryland and Virginia also had numerous large iron foundries, such as the "Onion Ironworks" in Maryland, which was outfitted with two forges, a furnace, sawmill, and grist mill and which sat on an eight-thousand-acre plot.

All of these enterprises operated within the constraints of empire. Authorities in London knew about Hasenclever; some even invested in his business.[17] As long as Hasenclever restricted himself to producing iron pigs or bars for export to the mother country, no mercantilist objected to his enterprise; other colonial ironmasters could expect the same sort of friendly treatment. Yet the American iron industry suffered many setbacks. Hasenclever went bankrupt after a few years because operation costs and labor problems had cut his profits to nothing. He was not the first nor was he the last to go under because of a shortage of capital and workers. Other ironworkers held on but could not expand or experiment with new techniques.[18] In addition, and more important to colonial agitators, imperial proscriptions—on paper, at least—limited the iron industry. While colonial ironmasters were encouraged to produce pig and bar iron, they were told not to make finished products. Pennsylvania ironworkers largely ignored the Iron Act of 1750 forbidding the erection of new steel or slitting and plating mills, but Parliamentary decrees that theoretically kept them out of finish work created grievances which "must be added to other fundamental causes of the Revolution."[19]

Using the iron industry as a case in point, some historians have blamed both the rise of colonial manufactures and their stunted growth on the British imperial system. If Parliament prodded ironmasters with incentives in the form of rewards and the promise of contracts for raw iron, it at the same time set an unfair ceiling on the industry by forbidding the processing of iron into steel or tools. Mercantilistic connections, assert historians critical of the empire, encouraged Americans to import more than they exported, thereby perpetuating an unfavorable balance of trade. Furthermore, lists of enumerated articles under successive navigation acts cut into the natural export trade by restricting commerce to the empire and blocking lucrative ties with "outsiders" in Europe and the West Indies. British policy caused insoluble problems because it was contradictory. It encouraged the production of some goods with bounties on one hand, such as hemp and pig iron, and forbade the making of fine apparel and steel on the other.[20] Some colonists fell back on small-scale home manufactures to recoup their losses and raise money to buy more expertly made British manufactures. According to critics of the empire, they had no alternative. Whatever economic gains they had made over the years, they had made despite, not because of, the imperial connection.

Other historians evaluate imperial relations in a more favorable light, emphasizing that restrictions were slight, given colonial conditions. Besides, until after 1763 the colonists violated parliamentary restrictions with virtual impunity. Those all-too-common failures and false starts in colonial industry could be more accurately ascribed to poor business practices, insufficient capital, or undeveloped markets. If not for the reciprocal nature of the empire, which minimized mercantilistic restrictions through privileged trade with Britain and its other colonies, the Americans would have fared much worse.[21]

The comparative restrictiveness or reciprocity of the British empire, whether in theory or practice or both, should not obscure a more important point. Frustrated colonists believed that imperial policy was exploitative. Their perception determined their behavior, and their view eventually held sway. They were not content with their level of economic activity, even if they did enjoy a relatively high standard of living and a growing per capita income. They blamed the empire for whatever problems they detected within the colonial economy, and historians who have since criticized the empire simply echo their complaints. For instance, a patriot calling himself "A North American" stated "that the Mother Country has herself" fomented the rise of home manufactures with policies dating from the Sugar Act in 1764. "While the Mother Country permitted the Colonies to carry on a trade with the French and Spanish settlements," he contended, they "never attempted to set up any Manufactories" running counter to British interests. Americans were more naturally inclined toward agriculture because of the newness of society, the availability of land, the shortage of skilled workers, and the lack of industrial skills. But the threat of economic strangulation, he concluded, moved them to promote manufacturing.[22] A petition signed by over two hundred Philadelphia merchants in 1766 was couched in similar terms, accusing Parliament of having forced Americans to "set up Manufactories of their own."[23]

A few colonists tried to counter these arguments. They believed unrealistically that American manufactures would actually strengthen the weakening bond between the mother country and her "plantations," acting as the "cement of empire." John Mascarene was one such believer, so he advocated potash manufacture to increase the "mutual Advantage" of trade. Since the most frequent complaint expressed by his fellow New Englanders was that specie flowed to Britain with none returning, Mascarene argued that potash manufacture for export would check that drain. America's vast woodlands were an endless source of potash, he asserted; landholders would benefit, potash makers would benefit, and Britain would benefit by purchasing potash from the colonies instead of Continental Europe.[24] An Englishman of the same opinion suggested that the wine culture be transplanted to the South. "It is certainly our Duty here in England, to give" the colonies "all possible Encouragement to set up such Manufactories as will not interfere with our foreign Trade," he advised.[25] Basing his conclusions on a trip to America made in 1744, Alexander Cluny argued that colonial subsistence production of wool, hemp, and flax would not undermine British markets and would reassure Americans that their liberties were not endangered. Colonial manufactures for colonial consumption would prevent "idleness" which bred dissatisfaction with the empire and would also divert New England from the carrying trade, that constant source of friction.[26] Daniel Dulany's influential pamphlet condemning the Stamp Act of 1765 followed the same train of thought. "Let the Manufactures of *America* be the Symbol of Dignity, the badge of Virtue," he wrote, "and it will soon break the Fetters of Distress." Self-sufficient production of simple clothing and foodstuffs would strike a blow at British imperial policy and force a quick remedy to colonial economic ills.[27]

In the long run, however, home manufactures could not possibly act as the "cement of empire," regardless of the schemes of Mascarene and admonitions of Dulany. On the contrary, manufactures assumed a symbolic importance leading ultimately to the opposite role. Recognizing that the colonists had to be self-sufficient in certain basic articles, colonial legislatures had traditionally promoted some local household industry. Parliament and the Board of Trade did not object. They tolerated colonial legislation promoting selected industries and even encouraged the legislatures to support certain trades and industries, such as the Hasenclever ironworks and Lynn cordwainers. Home manufactures, then, had been a part of the empire for years. But the meaning of home manufactures changed over time, because of both natural impulses and imperial legislation. The early emphasis of colonial legislatures on a few rudimentary household goods came to be overshadowed by the emphasis on more competitive manufactures. That shift in emphasis disturbed mercantilists in London.[28] What historian William Weeden called a "crude social force" by the 1750s had not yet evolved into a quest for economic independence, but Americans moved inexorably in that direction. There was more to Benjamin Franklin's advice to "manufacture as much as possible and say nothing" than meets the eye.[29]

Daniel Dulany and others like him played unwittingly into the hands of the more radical imperial critics. Dulany did not want to dismember the empire, yet he did want to improve the colonial economy. Accordingly, he supported the home manufactures movement that sprang up in the 1760s. In doing so he and dozens of others who remained loyalists during the War of Independence helped push the more independent-minded into power. Just as the revolutionaries escalated their political demands through the 1770s, they extended their arguments to cover the need for economic—and technological—independence. They eventually tied their technological frustrations to their growing political aspirations.

Agitation for home manufactures became a secular jeremiad. Merchants and mechanics alike were urged to shun ostentatious European finery, wear homespun, and buy American-made goods. The home manufactures movement sprang from a growing if still embryonic belief in the power of technology to uplift and transform society.[30] Leading patriots viewed improved handicraft skills, more extensive manufacturing, and progressive agriculture as the elixir for an ailing economy. Advocates of change linked improved farming—new plant strains, new tools, and better land use—with manufactures and commerce. They hoped that the latest tools and ideas would lift their economy out of the doldrums. Manufactures in particular, they concluded, would have an almost utopian effect. Improved manufactures would achieve everything from a reversal of the perennial balance of trade deficit to keeping the poor out of mischief by putting them to work in factories.

From this broad-scale self-improvement program, colonial technological enthusiasts labored to produce a better society. Therein lay the rub, for that ideal society could not be reconciled with imperial arrangements. Many colonists rejected the imperial justification for such things as the Iron Act of 1750 and sought to manufacture whatever they wanted. Just as important, by emphasizing a political solution to their economic woes, they did not see the technological problems lurking behind the scenes. Those problems would have been there regardless of the politics of empire. Thus the iron industry was kept small as much if not more by the state of colonial society as by imperial legislation. So too with other trades that fell within the restrictions of the navigation acts.

The home manufactures movement consequently took on a peculiar character. Sometimes it appeared to be quite strong, and at other times it seemed to be so weak it was about to expire. Outwardly, at least, the movement ebbed and flowed in reaction to British imperial policy after 1763. To most colonial agitators the navigation system until then had been "irritating" rather than "onerous," a source of friction, but friction kept to a minimum by lax enforcement.[31] By 1763 they had become more worried and imputed malicious motives to British actions. Consequently, the home manufactures movement that had begun in the 1750s did not really gain momentum until the 1760s, particularly after passage of the Stamp Act in 1765 and Townshend duties in 1767. After stalling it picked up again at the end of 1774 with the formation of the Continental Association and peaked in emotional intensity during the early part of 1775, when war loomed just over the horizon. Non-importation and eventually non-intercourse, from the Stamp Act to the Coercive Acts in 1774, went hand in hand with home manufactures.

The colonies had long had the potential for just such a movement. New England had been subject to "spinning crazes" since the 1720s, and in Boston a Society for Encouraging Industry and Employing the Poor, organized in 1751, helped bring on a new manufacturing frenzy.[32] Still, the factory erected by the society experienced difficulty from the start. Champions of home manufactures tried to awaken Bostonians to the need for patronizing it. They argued that patronage would bring jobs for the poor and "our Land will be further improved, our Importation lessen'd and our Exportation increased."[33] Salem Congregational minister Thomas Barnard spoke to the society in 1758, preaching the need for industriousness and industry. He contended, as would others around the colonies, that the lure of western lands impeded manufacturing progress by drawing off the population. Since westward migration took craftsmen away from their shops to farm, Barnard implied that it had an injurious effect on technology in general. He also stressed that the Puritan fathers had been very concerned with the promotion of agriculture and manufactures, hoping thereby to give home manufactures the sanction of history. Even more important, he outlined two activities that became common among colonial manufacturing societies throughout the 1760s and early 1770s. First, he insisted that premiums should be awarded for experiments in new ways of planting flax and hemp or for new modes of manufacturing. Second, he reminded society members of the need to become familiar with the latest discoveries in Europe.[34] Barnard proposed a direct connection between British and American industry, with British industrial skill paving the way for American expertise. Barnard saw the transit of technology—people, machines, and ideas—as the key to progress.

His recommendations and those of other manufacturing champions did not go unnoticed. Over the next decade more and more Bay Colony residents pledged themselves to the support of home manufactures. In 1764 fifty Boston merchants agreed to forgo imported lace and ruffles, while mechanics agreed to wear only Massachusetts-made leather aprons. At a mass meeting held in Faneuil Hall in October 1767 the freeholders and "other inhabitants of the Town of Boston" promised to "lessen the Use of Superfluities"—European luxury items—and buy more home manufactures instead.[35] Bostonians subsequently bragged that "millstones and grindstones, equal if not superior, to British, are now to be had among ourselves." The *Boston Gazette* was published on paper made from linen rags brought in by the townsfolk. Patriots claimed that the "spirit of the people" in "all of the colonies to invent and promote manufactures" was such that their "distressed and burthened continent" would soon be saved.[36]

Rhode Island and Connecticut also caught the home manufacturing mood. Back in 1752 Ezra Stiles had contended that Rhode Island could easily clothe itself. After all, stressed Stiles, the colony had nearly twice as many sheep as it had people.[37] Providence women, like women in Boston and other New England towns, held marathon spinning bees. In November and again in December 1767, town meetings in Providence and Newport voted unanimously to support home manufactures to lift the colony out of debt. A non-importation resolution was passed, along with resolutions to promote linen and wool production.[38] Connecticut towns followed suit, one resident asking not for "Revolution—God forbid it; but . . . a steady Perseverance in Manufacturing our raw Materials, for our own Consumption."[39] New Englanders were everywhere exhorted with a religious fervor to cleanse themselves of corrupt European influences and buy American.

At a Meeting of the Freeholders and other Inhabitants of the Town of *Boſton*, legally aſſembled at *Faneuil*-Hall, on Wedneſday the 28th of *October*, 1767.

THE Town then took into Conſideration the Petition of a Number of Inhabitants, "That ſome effectual Meaſures might be "agreed upon to promote Induſtry, Oeconomy, and Manufactures; thereby "to prevent the unneceſſary Importation of European Commodities, which threaten the Country "with Poverty and Ruin:" Whereupon in a very large and full Meeting, the following Votes and Reſolutions were paſſed Unanimouſly.

Whereas the exceſſive Uſe of foreign Superfluities is the chief Cauſe of the preſent diſtreſſed State of this Town, as it is thereby drained of its Money; which Misfortune is likely to be increaſed by Means of the late additional Burthens and Impoſitions on the Trade of the Province, which threaten the Country with Poverty and Ruin:

Therefore, *VOTED*, That this Town will take all prudent and legal Meaſures to encourage the Produce and Manufactures of this Province, and to leſſen the Uſe of Superfluities, & particularly the following enumerated Articles imported from Abroad, viz. *Loaf Sugar, Cordage, Anchors, Coaches, Chaiſes and Carriages of all Sorts, Horſe Furniture, Men and Womens Hatts, Mens and Womens Apparel ready made, Houſhold Furniture, Gloves, Mens and Womens Shoes, Sole-Leather, Sheathing and Deck Nails, Gold and Silver and Thread Lace of all Sorts, Gold and Silver Buttons, Wrought Plate of all Sorts, Diamond, Stone and Paſte Ware, Snuff, Muſtard, Clocks and Watches, Silverſmiths and Jewellers Ware, Broad Cloths that coſt above* 10s. *per Yard, Muffs Furrs and Tippets, and all Sorts of Millenary Ware, Starch, Womens and Childrens Stays, Fire Engines, China Ware, Silk and Cotton Velvets, Gauze, Pewterers hollow Ware, Linſeed Oyl, Glue, Lawns, Cambricks, Silks of all Kinds for Garments, Malt Liquors and Cheeſe.*——And that a Subſcription for this End be and hereby is recommended to the ſeveral Inhabitants and Houſholders of the Town; and that *John Rowe*, Eſq; Mr. *William Greenleaſe*, *Melatiah Bourne*, Eſq; Mr. *Samuel Auſtin*, Mr. *Edward Payne*, Mr. *Edmund Quincy*, Tertius, *John Ruddock*, Eſq; *Jonathan Williams*, Eſq; *Joſhua Henſhaw*, Eſq; Mr. *Henderſon Inches*, Mr. *Solomon Davis*, *Joſhua Winſlow*, Eſq; and *Thomas Cuſhing*, Eſq; be a Committee to prepare a Form for Subſcription, to report the ſame as ſoon as poſſible; and alſo to procure Subſcriptions to the ſame.

And whereas it is the Opinion of this Town, that divers new Manufactures may be ſet up in America, *to its great Advantage, and ſome others carried to a greater Extent, particularly thoſe of Glaſs & Paper*

Therefore, *Voted*, That this Town will by all prudent Ways and Means, encourage the Uſe and Conſumption of Glaſs and Paper, made in any of the Britiſh American Colonies; and more eſpecially in this Province.

[*Then the Meeting adjourn'd till* 3 *o'Clock Afternoon.*]

III o'Clock, *P. M.*

THE Committee appointed in the Forenoon, to prepare a Form for Subſcription, reported as follows.

WHEREAS this Province labours under a heavy Debt, incurred in the Courſe of the late War; and the Inhabitants by this Means muſt be for ſome Time ſubject to very burthenſome Taxes:——And as our Trade has for ſome Years been on the decline, and is now particularly under great Embarraſſments, and burthened with heavy Impoſitions, our Medium very ſcarce, and the Balance of Trade greatly againſt this Country:

WE therefore the Subſcribers, being ſenſible that it is abſolutely neceſſary, in Order to extricate us out of theſe embarraſſed and diſtreſſed Circumſtances, to promote Induſtry, Oeconomy and Manufactures among ourſelves, and by this Means prevent the unneceſſary Importation of European Commodities, the exceſſive Uſe of which threatens the Country with Poverty and Ruin—DO promiſe and engage, to and with each other, that we will encourage the Uſe and Conſumption of all Articles manufactured in any of the Britiſh American Colonies, and more eſpecially in this Province; and that we will not, from and after the 31ſt of *December* next enſuing, purchaſe any of the following Articles, imported from Abroad, viz. *Loaf Sugar*, and all the other Articles enumerated above.—

And we further agree ſtrictly to adhere to the late Regulation reſpecting Funerals, and will not uſe any Gloves but what are Manufactured here, nor procure any new Garments upon ſuch an Occaſion, but what ſhall be abſolutely neceſſary.

The above Report having been conſidered, the Queſtion was put, Whether the ſame ſhall be accepted? *Voted unanimouſly in the Affirmative.*—And that ſaid Committee be deſired to uſe their beſt Endeavours to get the Subſcription Papers filled up as ſoon as may be. Alſo, *Voted unanimouſly*, That the foregoing Vote and Form of a Subſcription relative to the enumerated Articles, be immediately Publiſhed; and that the Selectmen be directed to diſtribute a proper Number of them among the Freeholders of this Town; and to forward a Copy of the ſame to the Select-Men of every Town in the Province; as alſo to the principal City or Town Officers of the chief Towns in the ſeveral Colonies on the Continent, as they may think proper.

Atteſt,

William Cooper, *Town-Clerk.*

Then the Meeting was Adjourn'd to the 20th Day of November next.

2. Boston home manufactures notice, pledging town residents to the support of local industry and a renunciation of "foreign Superfluities." Courtesy of the American Antiquarian Society.

The New York Society for the Promotion of Arts, Agriculture, and Economy was organized in 1764 and dedicated to the same goals as the myriad New England associations and clubs. The society established a linen manufactory run under a putting out system. Spinners picked up raw flax and returned yarn, which other workers wove into cloth.[40] The society also sponsored special market days for home manufactures where people could purchase shoes, stockings, gloves, and cloth made by local craftsmen. Premiums put up by the society helped make the twice monthly sales a success. These premiums, awarded annually, offered sums from £3 to £20 for shoes, potash, stockings, looms, flax mills and even for teaching spinning.[41] New York craftsmen sometimes attended society meetings where they displayed everything from shovels and scythes to paper hangings. All New Yorkers, like New Englanders and Pennsylvanians, were asked not to eat lamb in order to increase the amount of wool available in the colonies. They were also educated in the baneful effects of European imports on the pocketbook and the soul.[42]

On December 12, 1765, the *Pennsylvania Gazette* informed its readers that a "Market for all Kinds of HOME MANUFACTURED GOODS" would be opened in Philadelphia by William Smith three days a week. Smith acted as a middleman, selling to the public any and all homemade articles brought in by their makers. A number of stocking factories shot up, with Thomas Bond running one in his home by 1768. Businesses that had been around for years changed their advertisements to include a statement about patriotism and buying American, trying, it would appear, to capitalize on the excitement of the moment and attract more customers.[43]

The drive among Pennsylvanians for greater production brought dividends to agriculture as well as industry. From April 1765 to April 1766, Philadelphia merchants exported 357,522 bushels of wheat, 198,516 barrels of flour, 60,206 bushels of corn, 882 tons of pig iron, and 813 tons of bar iron, figures well above those of earlier years.[44] By 1770 Lancaster County reportedly had between five hundred and seven hundred spinning wheels and about fifty looms. "So great is the Spirit of Homespun among our good Females," proclaimed Lancaster resident William Atlee, that scarcely a household lacked a spinning wheel.[45] An alarmed General Thomas Gage had earlier written to Lord Shelburne, the secretary for colonial affairs, "I could not help being surprised at the great Increase" in Philadelphia "in Buildings, Mechanicks and Manufactures." Unlike some colonists, noted Gage, Pennsylvanians eschewed grandiose schemes for practical, moneymaking enterprises. "If they go on as they have hitherto done," he continued, "they will probably in a few years Supply themselves with many Necessary Articles, which they now import from Great Britain."[46]

Sure enough, Pennsylvanians were flagrantly violating Parliamentary laws proscribing the manufacture of certain items, particularly steel. The Iron Act, like many other pieces of unpopular legislation, was virtually ignored. Thirty-four Philadelphia craftsmen signed a testimonial that steel made by the local firm of Humphreys and Zane proved "EQUAL, if not superior, to any imported from England." That Humphreys and Zane did so illegally by violating the Iron Act did not seem to bother them. Humphreys had in fact been granted £100 by the Pennsylvania General Assembly to carry on his work, while Zane continued to produce steel at Trenton, New Jersey, through 1772.[47] By that same year Philadelphia had a china factory, a glass factory, a paper factory, and a pot and pearl ash business.[48]

The Southern colonies, though lacking the basic industries and class of craftsmen common in the Middle and Northern colonies, vowed nonetheless to support home manufactures.[49] In 1757 Virginia had awarded an Englishman, Thomas Stephens, £100 to procure materials for a potash furnace in Williamsburg. The House of Burgesses eagerly accepted Stephens's offer to instruct Virginians in the art of potash-making.[50] In 1759 Burgesses went on to pass an act "for encouraging Arts and Manufactures," granting premiums up to £20, raised to £100 in 1762, for "useful discoveries."[51] The preceding action was motivated primarily by concern for Virginia's faltering economy during the French and Indian War, just as later legislation resulted from concern over imperial duties. Thus, in 1769 and again in 1770, Burgesses passed bills providing premiums for hemp with the hope that hemp would subsequently become a major export.[52]

John Wily, part owner of a fulling mill in Caroline County, emerged as Virginia's leading authority on hemp and home manufactures in general. "For as we have got in Debt by our Indolence and Extravagancy," Wily berated his readers, "sure there is no better Method to retrieve ourselves than by our Industry and Frugality." In a concisely written manual he described the process for dyeing, bleaching, and milling cloth and diagrammed the parts and function of a fulling mill and mechanical devices needed to work flax and card wool.[53] Though at first merely advocating the manufacture of coarse woolens and linen, by 1769 he had begun encouraging his fellow Virginians to make fine quality wool garments as well.[54]

Judging from the general public notice, home manufactures fever subsided after 1770 and became dormant. It did not revive until late 1774 with the outrage over a spate of imperial legislation that colonial agitators contemptuously labeled the Coercive Acts. The Boston Port Bill in particular raised patriot hackles. Provincial assemblies and committees of safety, this time following the lead of the newly formed Continental Congress, reaffirmed their dedication to home manufactures.

On October 20, 1774, Congress announced the formation of the Continental Association, reflecting its decision "that a non-importation, non-consumption, and non-exportation agreement" would speedily redress grievances. Promotion of "agriculture, arts and the manufactures of this country" was coupled with non-intercourse. Taking a stance similar to colonial assemblies in the 1760s, the association condemned the wearing of finery and exorbitant prices for homemade goods and recommended that sheep be spared for their wool, the wool to be made into cloth.[55]

Assemblies and committees of safety rallied to the call for economic nationalism and enacted legislation for the program envisioned by Congress. Resolutions adopted by Massachusetts were representative of those adopted elsewhere. The Provincial Congress there recommended that homemade goods be used in preference to imports; that hemp and flax be planted in larger quantities; that paper, glass, wool, and salt production be encouraged; that a manufactory for wool cards be erected; and that a society for promoting arts and manufactures be organized. In an ominous departure from earlier resolves, provisions were also made for a war industry. Gunsmiths were urged to make firearms and powderhorns, and production of saltpeter (for gunpowder) and steel was emphasized.

Virginia also backed manufactures "necessary for subsistence, clothing and defense."[56] Individual Virginia counties offered premiums, with James County promising £40 for the first five hundred wool cards, and Essex County pledging £50 for the first five hundred pairs of stockings. Chowan County, North Carolina, gave bounties for wool and cotton cards, wool and linen cloth, and steel.[57] A Marylander advised all of the colonies to concentrate on textiles, "notwithstanding the most perfect harmony should be again restored between Great Britain and America." Wealthy landholders, he added, should put up bounties and premiums, as was done in Scotland and Ireland, because of the "natural conexion" of agriculture and manufactures.[58]

The Pennsylvania convention mirrored Massachusetts and Virginia in backing steelmaking, firearms, and gunpowder production as well as the manufacture of glass, textiles, and paper. It also offered premiums to individuals introducing "new machines or techniques of production into the state." Many of the machines entered in award competitions were for spinning and weaving cloth. In fact, much of the home manufactures movement hinged on textiles because textiles were looked to as the great test of technological independence. The colonies had been producing coarse cloth for decades, but that cloth could not compete with cloth imported from the mother country. The enthusiasts wanted to become more competitive, so they tried to introduce a combination of cottage and factory work.[59]

Pennsylvanians initiated the most ambitious textiles manufacturing scheme to be seen in America before the War of Independence. In February 1775 a group of them banded together and formed the United Company of Philadelphia for Promoting American Manufactures. Dr. Benjamin Rush was the moving force behind the company and its first president. Rush had been an advocate of home manufactures for some time. In 1768 he expressed an interest in manufacturing and a year later stated, "There is but one expedient left whereby we can save our sinking country, and that is by encouraging American manufactures." A key to lasting success would be to invite "hundreds of artificers of every kind" to "come over from England and settle among us." That same year Rush visited London and may well have attended meetings of the prestigious Royal Society for the Encouragement of Arts, Manufactures and Commerce. In 1772 he became a corresponding member of that society.[60]

On March 16, 1775, Rush addressed members of the United Company who had assembled in Carpenters' Hall. He stressed that textiles were essential because non-importation cut off the flow from Europe and reduced the profits of British merchants. Home manufactures could teach Britain the error of its ways. "Manufactures, next to agriculture, are the basis of the riches of every country," he told his listeners. Bringing manufactures to America, the "land of liberty and plenty," would raise the textiles industry from the "torpid" exploitative state into which it had slipped in Europe. A combination of home spinning and weaving and factory finish work would employ the poor, invigorate the economy, and bring about self-sufficiency. Rush warned that "a people . . . dependent upon foreigners for food or clothes, must always be subject to them."[61] That the British fell into Rush's "foreigners" category seems likely. Rush had admonished an audience a year earlier that mechanization must be selective, with the colonies adopting only those manufactures that would raise revenues without imperiling the lives and health of the people. The British, so far as Rush was concerned, had made manufactures the enemy rather than the friend of the people. Americans must avoid repeating the mistake.

Subscriptions to the United Company sold rapidly, at £10 apiece. Half of the money went to erecting a factory at Market and Ninth Streets and half to paying twelve managers to run it. The company recruited flax dressers, spinners, and weavers and soon boasted a labor force of four hundred. It worked through a putting out system similar to that of the New York manufacturing experiment of 1764. Local women were requested to bring their yarn to the factory for weaving, finishing, and eventual sale as cloth bolts. During the first few months of operation, the United Company actually paid dividends to its investors and managed to survive until the British occupation of Philadelphia in 1777.[62]

Nevertheless, despite the initial enthusiasm among Philadelphians, the company was always crippled by a shortage of capital, experienced factory workers, and long-term investment. The United Company, though an important beginning, could not by itself reverse the fortunes of colonial textiles. It suffered from the same problem as every other overambitious colonial manufacturing venture.

A 1774 prophecy that colonial dependence on the mother country "will, in a few years be reversed, and like the storks, the parent owe her subsistence to her children" had been premature.[63] Despite the zealousness of home manufactures enthusiasts, America had not become "industrialized" in any sense of the word by 1775. That this would be the case had been foreseen by a few of the better informed. A Pennsylvanian scolded his fellows in 1768 for only promoting home manufactures during times of imperial crisis. For home manufactures to have any prolonged success, he pointed out, support of them had to be constant.[64] Indeed, widespread, sustained backing for non-importation and financing of manufactures had been difficult to sustain. Support fluctuated over the years. Even at the height of the movement numerous merchants ignored non-importation pledges and imported European cargoes. Businessmen in Philadelphia, Boston, and New York chose to trade rather than practice self-denial. They may have pledged their support to home manufactures, but such pledges were easily—and often—broken.

Historian Arthur Schlesinger argued consequently that the colonists "were interested only in tiding over a difficult period and not in laying the foundations of permanent industries."[65] Lending credence to Schlesinger's thesis, a 1768 report out of Boston stated that the "publick spirit" for manufactures had fallen off, "the adventurers having met with nothing but disappointments in procuring materials, scarcity of money, and high prices of wages."[66] Rather than persevere they invested in the carrying trade, an established source of income.

Schlesinger's conclusion that the entire movement was episodic and temporary is only superficially correct. Though the movement gained headlines intermittently and seemed to run in spurts, it was part of a deep social undercurrent. This is not to say that the production index for home manufactures rose steadily, because it obviously did not. But slow production growth can be attributed as much to entrepreneurial inexperience and undeveloped markets as to lack of consistent interest. Schlesinger's point that the whole concept of self-sufficiency was unrealistic, a wild punch at a British bogeyman, is well taken. Americans unquestionably rallied behind home manufactures in a crisis more than during a calm, hence the episodic appearance. Schlesinger notwithstanding, the colonists knew manufacturing was essential to economic maturation and to any long-term political gain, both of which they sought. The seriousness of imperial contentions in 1765, 1767, and 1774 compelled many of them to invest in high-risk ventures. When tensions eased, they naturally went back to more settled pursuits. Inconsistent support does not necessarily mean that they did not want to lay the "foundations of permanent industries"; rather, individual merchants did not want to badly enough to chance a substantial and possibly paralyzing financial loss. In addition, as Schlesinger himself emphasized, most merchants were not all that eager for political independence. There is no reason to expect them to have pushed for complete technological independence until they had reconciled themselves to political independence. Furthermore, the lack of certain machines, people skilled in their use, and basic industrial talents kept them from laying those "foundations."

Technological deficiencies, it could be contended, contributed as much to the limited achievements of the home manufactures movement as the absence of ardent desire. Production rose and fell, hope changed to despair, and vice versa, because agitation in the 1760s and 1770s took manufactures beyond the point where they could operate profitably. The enthusiasts built their enterprises on a shaky, emotional foundation, instead of on a sound economic base. American patriots had tried to condense a major social movement into a decade. They expanded their technological objectives as they increased their political demands. As a consequence, they jumped ahead of their technological expertise and the flexibility of a limited market. Yet they did not, and probably could not, understand this at the time. Only after the political tie had been severed and they were on their own would they discover the complexities of technological retooling. For the time being they could only blame the British for their failures.

The home manufactures movement had burst forth in a handicraft society. The type of industrialization advocated by manufacturing enthusiasts was entirely new to a people accustomed to localized, small-scale production. The colonists, including the enthusiasts, had no experience with the new machines and new production techniques they advocated. They had neither the investment capital nor the entrepreneurial expertise to pursue such heady experiments with much hope of success. They had been unrealistically sanguine.

Not surprisingly, the manufacturing societies fell short of their goal. Businesses begun under their auspices collapsed, left and right. The linen manufactory set up in Boston went from a production peak in January 1766 to bankruptcy the following November. The New York operation also fell by the wayside after a short while. A glass factory erected in Lancaster County, Pennsylvania, by Henry Stiegel in 1771 suffered a similar fate. Stiegel brought in workmen from Germany and held lotteries to stay open, but his luck ran out in 1774. The china factory opened in Philadelphia with great fanfare in 1771 was put up for auction a year later.[67] A New York law granting bounties for hemp was stillborn from lack of support, while a Virginia House of Burgesses bounty on textiles never made it out of committee.[68]

Reports on the limitations and relative scarcity of colonial manufactures outnumbered reports of success. In accordance with the wishes of George III, in 1766 Parliament had issued a circular letter ordering the governors of all the American colonies to report on the state of manufactures within their borders. The subsequent responses echoed one another: manufacturing did not thrive in the colonies.[69]

Benjamin Franklin wrote in March 1768 that Parliament had received statements on the condition of manufactures from all of the colonial governors except two. "They are all much in the same strain, that there are no manufactures of any consequence," Franklin sighed. Colonial textiles had not advanced beyond the making of coarse linens in most areas; in the Southern colonies a vast majority wore English apparel even though many households spun and wove some coarse cloth. Of all the governors submitting their findings, Franklin noted, only the governor of North Carolina reported measurable progress, and that was in lumbermilling. Indeed, the province had fifty sawmills on one river alone. Nevertheless, those sawmills had come in response to demand in British and West Indies markets, not home manufactures patriotism.[70] Lieutenant Governor Francis Fauquier of Virginia wrote that "there is but one Manufactory of the least importance" in his colony, that being the making of pig and bar iron.[71] Governor Francis Bernard of Massachusetts assured Parliament that "for New England to threaten the mother country with manufactures is the idlest bully that was ever attempted to be imposed upon sensible people." Bostonians denied the veracity of Bernard's slighting remark, but not very convincingly.[72] Governor Henry Moore of New York conceded that many families in his province had looms and made coarse linen cloth, but other manufactured products came from England. Most craftsmen abandoned their trade to farm, concluded Moore, because of the plenitude of land and the restricted market for home manufactures.[73] Governor John Wentworth of New Hampshire reported in March 1768, and again in 1771, and his findings confirmed those of Bernard and Moore. Pot and pearl ash production did well, but otherwise "every mechanic branch has diminished" for the same reasons given by Moore.[74] Reports from the South echoed those sentiments. Comparable accounts on the state of American manufactures circulated just before the war.[75]

Nonetheless, the home manufactures movement should not be dismissed as an abject failure. It did serve to raise the industrial consciousness of Americans, helping them to make the mental connection between technological and political independence. They began to look to technological innovation as a source of socioeconomic change, a way of promoting progress. Investment in non-agricultural pursuits was common by 1775, and not just in shoemaking and overseas commerce. Furthermore, most colonies became somewhat less dependent on imports. "In one field after another," James Henretta has concluded, "the northern settlers of America were achieving economic self-sufficiency through the creation of import substituting industries."[76] The colonists proved to be remarkably adept at extractive industries such as fishing and lumbering. They had also surged ahead in related areas like shipbuilding and in small-scale handicrafts.

These successes notwithstanding, complete economic self-sufficiency had not been attained by 1775. This lack of self-sufficiency would cause considerable difficulty during the war, particularly since some patriots could not distinguish between rhetoric and reality. Their goal of self-sufficiency was unrealistic; belief that they had reached that goal was even more so. Localized production may have been adequate when the colonists remained content to be a part of the empire, but it was found wanting once their nationalistic ambitions had been stirred. Their desire to erase the stigma of inferiority derived from mercantilistic economic relations caused them to search for the means to improve their manufactures. Noting modest pre-war advances in the iron, potash, textile, and flour milling industries, and the money brought in from grain exports, they assumed that they could produce enough of whatever they needed. They were sadly mistaken. In 1775 the handicraft industry of Colonial America was not ready to supply a nation at war.

Moreover, the financial success of an ironmaker here or a shipwright there, while strengthening the economy, had done relatively little to change the technological complexion of colonial society. The colonists had by 1775 duplicated, on a smaller, perhaps more fragmented, scale, most of the handicraft trades available in the mother country. That they trailed in a few areas is attributable to both the economic restrictions of mercantilism and the youth of colonial society. Colonial America had not been technologically backward. At the same time, it did not have the technologies to sustain what became the ambitions of a people no longer content to live in the empire. When some colonists attempted to promote modes of production new even to the mother country, they found that their aspirations outran their capabilities. During the war the patriots did expand some traditional handicraft trades, despite the lingering provincialism of colonial politics and an unrelieved shortage of capital. They could not, however, push on to experiment with new machines and new methods of production.

The War of Independence temporarily cut off contact between manufacturing enthusiasts in the colonies and the sources of innovation in Great Britain. Severed ties hurt Americans dreaming of a more vibrant economy. Throughout the Colonial Era they had depended on a transit of technology for the successful assimilation of handicraft trades and small-scale manufactures. Whatever self-sufficiency they had achieved by 1775 came more from the adaptation of traditional modes of work and tools to the environment than from the implementation of new devices they had invented. The success or failure of some industries set up just before the war depended, ironically, on the continued importation of people and machines from Great Britain. As a case in point, while Charleston, South Carolina, residents pledged to support nonimportation in 1769, they carefully exempted from the list of outlawed goods wool cards, millstones, steel, gunpowder, saltpeter, firearms, and assorted tools.[77] They were more concerned with getting the materials they needed to pursue their economic and political goals than they were with being consistent.

Americans depended on the mother country for technological information of all sorts before and after the War of Independence, information on methods of business organization and conduct as well as production techniques. Thomas Barnard, Benjamin Rush, and others realized that American craftsmen were isolated from the mainstream of innovative thinking in Europe. They could not by themselves bring about an industrial upsurge in the colonies, particularly since so many left their shops to till the soil. The colonies needed the innovative ideas, skills, and technical expertise of immigrant craftsmen. They needed access to British journals and the findings of scientific societies. Because those ideas were screened and some craftsmen could not emigrate, the colonists were only dimly aware of certain key ingredients necessary for the changes they wanted to make.

The War of Independence came at a most inopportune moment, catching the colonists mid-stride between abject dependence on Britain and their own drive for economic self-determination. Life in Colonial America had had its technological advantages and disadvantages, the sheer expanse of arable land and wealth of resources acting as both technological retardant and stimulant. To most farmers, innovative planting and soil conservation in such a lush setting seemed unnecessary at best, wasteful of time, energy, and money at worst. As a result, scientific farmers like George Washington could not easily change the habits of their neighbors. Neither could patriots who pushed for a united effort to counter British mercantilism. Manufacturing enthusiasts likewise found, to their chagrin, that they could not easily erect new industries. Patriotic public appeals did not produce many substantial gains, so advocates of change fell far short of their goals. Their technological aspirations had changed with their political aspirations, but their ability to bring rapid technological change had not.

The closer they moved toward political independence, the more pressure they put on themselves to build a more complex technological order, one where change could be accelerated and profits quickly realized. Advocates of technological change ran smack into political restrictions part and parcel of a mercantilistic empire. They attempted to launch an ambitious program of manufactures in an overwhelmingly agricultural society; they needed new machines, but they did not have the craftsmen to build them; they needed investment capital, but money was scarce; if colonial assemblies were willing to help, the government in London was not. Technological enthusiasts had little except an oversupply of desire.

The home manufactures movement had enjoyed boisterous if periodic backing, yet will to improve alone was not enough. Perhaps the boastful pride of technological enthusiasts masked a deeper insecurity. What those enthusiasts needed most they did not have: a technological capability to match their political ambitions. Economic historians have often observed that the technology of Colonial America was not dynamic enough to promote large-scale change. Thomas C. Cochran, for one, noted that "compared to later periods, technology in America from 1607 to 1783 was static."[78] Cochran, of course, did not mean that there was no change at all. The colonists were perfectly capable of supplying themselves with most basic necessities, and colonial sawyers, it should be remembered, used mill machinery superior to that used in the mother country. Nevertheless, the innumerable innovations made by colonial craftsmen did not transform the nature of production even if they did increase the mechanization of some trades. Millers and sawyers did not produce a pattern of innovation that could be imitated, a model that could be followed, an example that would stir the public.

Americans wanted to make a basic technological transformation. They spent the 1760s and early 1770s trying. The great visions of manufacturing enthusiasts before the War of Independence consequently remained visions—technological expressions of wishful thinking. American society had to change a good deal before the enthusiasts would see their dream become a reality. They needed more capital, more advanced machines, more craftsmen, more talented entrepreneurs, a political system set up to promote, not discourage, manufacturing experiments, and a general public more committed to technological change. Their conscious search for technological independence eventually produced the examples they needed, but those examples came after the war.

NOTES

1. James Boswell, *The Life of Samuel Johnson*, 4 vols. (London: T. Cadell and W. Davies, 1807), IV, 201.

2. Rodney C. Loehr, "Arthur Young and American Agriculture," *Agricultural History* 43 (1969):43-45; also see the contemporary observations of Samuel

Miller, *A Brief Retrospect of the Eighteenth Century,* 2 vols. (New York: T. & J. Sword, 1803), I, 375-389. T. S. Ashton, *The Industrial Revolution, 1760-1830* (London: Oxford University Press, 1948), p. 21, cautioned that the innovative Norfolk farming system was the product of many minds and hands but was by no means universally practiced in eighteenth-century England, even among the gentry.

3. Jared Eliot, *An Essay Upon Field Husbandry in New England* (New London: Timothy Green, 1748), part I, pp. ii-iii, and subsequent additions and printings through 1762, many of which were serialized in newspapers and magazines.

4. See for example the *Pennsylvania Gazette,* 24 March 1763, 20 September 1764, and 6 June 1771.

5. Merrill Jensen, "The American Revolution and American Agriculture," *Agricultural History* 43 (1969):107-124; George Washington in Paul L. Haworth, *George Washington, Country Gentleman* (Indianapolis: Bobbs Merrill Co., 1925), pp. 90-164. Duane Ball and Gary Walton, "Agricultural Productivity Change in Eighteenth Century Pennsylvania," *Journal of Economic History* 36 (1976):102-117, make a good case for the profitability and increased output of farms in Pennsylvania up to the Revolution. Nathan Rosenberg, "America's Rise to Woodworking Leadership," in Brooke Hindle, ed., *America's Wooden Age* (Tarrytown: Sleepy Hollow Press, 1975), pp. 37-62, pointed out that technological efficiency is relative. The colonists may have been ravenous in their use of the land, but in their case it was cheaper—and thus in a way more efficient—to use "resource-intensive" practices. James T. Lemon, *The Best Poor Man's Country* (Baltimore: Johns Hopkins University Press, 1972), pp. 150-151, 183, argues likewise. Pennsylvania farmers of the Revolutionary Era were content with their output, even if, by the standards of "scientific farmers," they did not use the best tools or adequately fertilize the soil. For Pennsylvania farmers interested primarily in making money, "extensive" farming made more sense than "intensive" farming.

6. *Virginia Gazette* (Rind), 19 July 1770; and ibid., 18 November 1773, for the same mill under a different owner.

7. Carl Bridenbaugh, *The Colonial Craftsman* (Chicago: University of Chicago Press, 1950), pp. 59-60. Descriptions of some mills can be found in the *Pa. Gazette,* 21 October 1762, 8 March 1764, 23 August 1764, and 4 December 1766. A description of a large New Jersey mill is in ibid., 19 May 1767. For mills in the breadbasket area along the Brandywine River near Wilmington, Delaware, see Peter Welsh, "The Brandywine Mills: A Chronicle of an Industry, 1762-1816," *Delaware History* 7 (1956):17-36.

8. *Pa. Gazette,* 3 February 1763 and 7 May 1767.

9. See Arthur Young's anonymously published *American Husbandry,* 2 vols. (London: J. Bew, 1775). Also James A. Henretta, *The Evolution of American Society, 1700-1815* (Lexington: D. C. Heath and Co., 1973), pp. 15-23, for New England; and Aubrey C. Land, "The Tobacco Staple and the Planter's Problems: Technology, Labor and Crops," *Agricultural History* 43 (1969):68-86, for the South.

10. Kenneth G. Davies, ed., *Documents of the American Revolution, 1770-1783,* 25 vols. (Shannon, Ireland: Irish University Press, 1972-; 21 volumes to date); IV (Calender), no. 1278, from PRO/CO, 5/395, fos. 41-42d. Davies drew his materials from the Colonial Office of the Public Record Office in London, hence future references will be PRO/CO, with appropriate volume and pages, followed by Davies's volume and pages. Davies has two types of volumes in his collection: calenders listing and briefly summarizing documents, and complete transcripts.

11. *Pa. Gazette,* 4 February 1768; and William B. Weeden, *Economic and Social History of New England, 1620-1789,* 2 vols. (Boston: Houghton Mifflin Co., 1890), II, 682-683. For a useful overview of colonial commercial and manufacturing life, see Evarts B. Greene, *The Revolutionary Generation, 1763-1790* (New York: Macmillan Co., 1943), pp. 35-66.

12. Bridenbaugh, *Colonial Craftsman,* pp. 41, 92.

13. For the iron industry see Arthur C. Bining, *Pennsylvania Iron Manufacture in the Eighteenth Century* (Harrisburg: Pa. Hist. Comm., 1938), pp. 11-34, 49-94; idem, *British Regulation of the Colonial Iron Industry* (Philadelphia: University of Pennsylvania, 1933); and John B. Pearse, *A Concise History of the Iron Manufacture of the American Colonies up to the Revolution and of Pennsylvania Until the Present Time* (Philadelphia: Allen, Lane, and Scott, 1876), pp. 7-90.

14. Pearse, *Concise History,* p. 47; *New York Gazetteer,* 16 February 1775; and the *New York Journal,* 22 January 1767, supplement. Lester J. Cappon, ed., *Atlas of Early American History, 1760-1790* (Princeton: Princeton University Press, 1976), p. 29, provides excellent maps plotting the location of ironworks throughout the colonies. The charts, graphs, and maps in this collection illustrate the dynamic economic life of the colonies, as does Alice Hanson Jones in her more involved *The Wealth of a Nation to Be* (New York: Columbia University Press, 1980). Revolutionary agitators nevertheless were not content.

15. The notice on quality can be found in the *Pennsylvania Journal and the Weekly Advertiser,* 11 April 1765; the market percentage was taken from Bridenbaugh, *Colonial Craftsman,* p. 61, who relied on figures given in Bining, *Pa. Iron Manfacture.*

16. Bining, *Pa. Iron Manufacture,* pp. 109-110; Pearse, *Concise History,* pp. 62-70. Also see Irene D. Neu, "The Iron Plantations of Colonial New York," *New York History* 33 (1952):3-24, for Hasenclever, the Sterling Works, and the Ancram Ironworks on Livingston manor. For the Principio and Baltimore ironworks, see Keach Johnson, "The Genesis of the Baltimore Ironworks," *Journal of Southern History* 19 (1953):157-179, which stresses that Maryland and Virginia actually led all the colonies in pig iron production well into the 1750s. Also see idem, "The Baltimore Company Seeks English Markets," *William and Mary Quarterly,* 3rd series, 16 (1959):37-60.

17. *Pa. Gazette,* 1 June 1769.

18. Ibid, 4 December 1766.

19. Bining, *Pa. Iron Manufacture,* p. 137.

20. For arguments on the restrictive nature of the empire, see Eleanor Lord, *Industrial Experiments in the British Colonies of North America* (1898; reprint ed., New York: Burt Franklin, 1969), pp. 124-126, 128-139; Rola M. Tryon, *Household Manufactures in the United States, 1640-1860* (Chicago: University of Chicago Press, 1917), pp. 17-18; and Curtis Nettles, "British Mercantilism and the Economic Development of the Thirteen Colonies," *Journal of Economic History* 12 (1952):105-114. Idem, "The Menace of Colonial Manufacturing, 1690-1720," *New England Quarterly* 4 (1931):230-269, outlines ill-fated British attempts to promote naval stores as a benign alternative to manufacturing in New England.

21. For more sympathetic descriptions of the empire, see Lawrence Henry Gipson, *The British Empire Before the American Revolution,* 15 vols. (New York: Alfred A. Knopf, 1958-1970), XIII, 181; Charles M. Andrews, *The Colonial Background of the American Revolution* (1924; revised ed., New Haven: Yale University

Press, 1931), p. 116; idem, *The Colonial Period in American History*, 4 vols. (New Haven: Yale University Press, 1934-1938), IV, 425n-428n. Gary M. Walton and James F. Shepherd, *The Economic Rise of Early America* (Cambridge: Cambridge University Press, 1979), pp. 96-112, also question the trade deficit argument. Edwin J. Perkins, *The Economy of Colonial America* (New York: Columbia University Press, 1980), p. 142, states flatly that the "degree of economic regulation and the level of imperial taxation were not significant causes of the War for Independence." I prefer the conclusions of Susan Previant Lee and Peter Passell, *A New Economic View of American History* (New York: W. W. Norton & Company, 1979), p. 39. Though Lee and Passell perhaps overemphasize the real possibility of future Parliamentary "exploitation," they are rightly sensitive to colonial paranoia regarding Parliamentary designs, a paranoia discussed fully in Bernard Bailyn, *The Ideological Origins of the American Revolution* (Cambridge: Harvard University Press, 1967). Adam Smith anticipated later historians like Bailyn when he noted in Edward Cannan, ed., *An Inquiry into the Nature and Causes of the Wealth of Nations* (New York: Modern Library, 1937), p. 549, that restrictions on manufactures struck some Americans as being "impertinent badges of slavery," disturbing evidence of Parliamentary malice.

22. *Pa. Gazette*, 27 June 1765.

23. From the Merchants of Philadelphia, in the Province of Pennsylvania, To the Merchants and Manufacturers of Great Britain, in MSS. Relating to Non-Importation Resolutions in Philadelphia, 1766-1775; American Philosophical Society Manuscripts (hereafter APS MSS.).

24. John Mascarene, *The Manufacture of Pot-Ash in the British North American Plantations Recommended* (Boston: Z. Fowle for T. Leverett, 1757), pp. 1-11. Mascarene later became a customs collector.

25. *Pa. Gazette*, 9 May 1765.

26. Alexander Cluny, *The American Traveller* (Philadelphia: Cruikshank and Collins, 1770), pp. 49-55. Others honestly believed that colonial manufactures could improve imperial relations—see the letter from Jonathan Trumbull to William Samuel Johnson of 29 January 1770 in the "Trumbull Papers," *Massachusetts Historical Society Collections*, 5th series, IX, 400.

27. Daniel Dulany, *Considerations on the Propriety of Imposing Taxes in the British Colonies* (Annapolis: Jonas Green, 1765), pp. 43-47.

28. Tryon, *Household Manufactures*, pp. 28-42; change in meaning on pp. 43, 104-111. For an example of an early act, see William W. Hening, ed., *The Statutes at Large*, 13 vols. (New York: R. & W. Bartow, 1823), I, 208.

29. Weeden, *Economic and Social History*, II, 679-681, 733-736; Franklin statement in Andrews, *Colonial Background*, p. 149. As Samuel Rezneck put it, "The will to manufacture was born before manufactures were started," in his "The Rise and Early Development of Industrial Consciousness in the United States, 1760-1830," *Journal of Economic and Business History* 4 (1932):785. Adam Smith, *The Wealth of Nations*, p. 360, contended that in the American colonies, as elsewhere, it was the "natural order" to pursue first agriculture, then manufactures, and finally foreign commerce. "Some of their lands must have been cultivated before any considerable towns could be established," wrote Smith, "and some sort of coarse industry of the manufacturing kind must have been carried on in those towns, before they could well think of employing themselves in foreign

commerce." While Smith's line of progressive development may have been oversimplified, he did hit on something in noting the tendency of Americans to want to broaden the range of their economic activity. And like some of the colonists themselves, he noted that "artificers" often abandoned their trades to farm, thereby slowing the rate of manufacturing growth and the very economic changes they desired.

30. Hugo A. Meier, "The Technological Concept in American Social History, 1750-1860" (Ph.D. dissertation, University of Wisconsin, 1950), pp. 535-538; other implications of the movement are discussed in Edmund S. Morgan, "The Puritan Ethic and the American Revolution," *WMQ*, 3rd series, 24 (1967):3-18.

31. For the "irritating" colonial system, see Vincent Harlow, *The Founding of the Second British Empire, 1763-1793* (London: Longmans, Green and Co., 1952), pp. 146-162. Also see Oliver M. Dickerson, *The Navigation Acts and the American Revolution* (Philadelphia: University of Pennsylvania Press, 1951), and the corrective to Dickerson in Thomas C. Barrow, *Trade and Empire* (Cambridge: Harvard University Press, 1967).

32. Tryon, *Household Manufactures*, pp. 85-87; William R. Bagnall, *The Textile Industry of the United States*, 2 vols. (Cambridge: Riverside Press, 1893), I, 29-50. Use of the poor in industry was not original. See Richard B. Morris, *Government and Labor in Early America* (New York: Columbia University Press, 1946), pp. 3-13. The best survey of the home manufactures movement in general is found in Arthur Schlesinger, *The Colonial Merchants and the American Revolution* (New York: Columbia University Press, 1918), although I do not agree with some of Schlesinger's conclusions. J. E. Crowley, *This Sheba, Self* (Baltimore: Johns Hopkins University Press, 1974), pp. 125-146, approaches the subject from a provocatively different angle.

33. Anonymous, *Industry and Frugality proposed as the Surest Means to make a Rich and Flourishing People* (Boston: Thomas Fleet, 1753), pp. 7-14; quote from p. 12.

34. Thomas Barnard, *A Sermon Preached in Boston, New England, Before the Society for Encouraging Industry, and Employing the Poor, September 20, 1758* (Boston: S. Kneeland, 1758), pp. 9-22.

35. *Va. Gazette* (Purdie and Dixon), 19 November 1767; and *Pa. Gazette*, 19 November 1767.

36. *Pa. Gazette*, 4 February and 11 February 1768.

37. Franklin B. Dexter, ed., *The Literary Diary of Ezra Stiles*, 3 vols. (New York: Charles Scribner's Sons, 1901), III, 360.

38. *Boston Gazette and Country Journal*, 7 April 1766; town resolves in the *New York Journal*, 17 December 1767.

39. Connecticut towns in the *Pa. Gazette*, 11 February 1768 and 8 December 1768; quote from the *Boston Gazette*, 6 March 1769.

40. Factory in the *New York Journal*, 27 November 1766. Notice of the society in the *Pa. Journal*, 3 January 1765; and the *Pa. Gazette*, 31 October and 12 December 1765; and 12 November 1767. General information on the society can be found in Brooke Hindle, "The Underside of the Learned Society in New York, 1754-1854," in Alexandra Oleson and Sanborn Brown, eds., *The Pursuit of Knowledge in the Early American Republic* (Baltimore: Johns Hopkins University Press, 1976), pp. 84-116.

41. *Va. Gazette* (Rind), 12 March 1767; *New York Journal,* 24 December 1766. Also see the *Pa. Journal,* 3 January 1765.

42. Lamb in the *New York Journal,* 13 April 1769; also see the *Pa. Gazette,* 19 February 1767, and a similar pledge by Pennsylvanians in the *Pa. Journal,* 11 April 1765; also see "A Linen Draper," *The Commercial Conduct of New York Considered* (New York, 1767), pp. 7-11.

43. *Pa. Gazette,* 29 September 1768, and 20 June and 26 December 1765; drumming up business also in ibid., 20 June 1765.

44. Ibid., 19 March 1767. Figures for New York are also given.

45. Ibid., 14 June 1770.

46. Clarence E. Carter, ed., *The Correspondence of General Thomas Gage,* 2 vols. (New Haven: Yale University Press, 1931-1933), I, 160-161.

47. *Pa. Gazette,* 4 January 1770; Bining, *Pa. Iron Manufacture,* pp. 154-155; also see the *Pa. Gazette,* 1 April 1772.

48. Ibid., 29 August and 26 September 1771, 19 August 1772.

49. For Annapolis see ibid., 4 January 1770; Charleston in ibid., 21 February 1765 and the *New York Journal,* 31 August 1769; Savannah in the *Pa. Gazette,* 9 November 1769.

50. H. R. McIlwaine and J. P. Kennedy, eds., *Journals of the House of Burgesses, 1619-1776,* 13 vols. (Richmond: Everett Waddey Co., 1915), VIII, 423, 492; *Va. Gazette* (Hunter), 22 April 1757; and Stephen's pamphlet the *Method and Plain Process for Making Pot-Ash Equal if not Superior to the Best Foreign Pot-Ash* (Boston: Edes and Gill, 1755).

51. McIlwaine and Kennedy, eds., *Journals of Burgesses,* IX, 109, 120, 124, 128; Hening, ed., *Statutes,* VII, 289-290, 563-566.

52. Hening, ed., *Statutes,* VIII, 363-364; McIlwaine and Kennedy, eds., *Journals of Burgesses,* XII, 13, 59, 62, 93, 94, 98.

53. John Wily, *A Treatise on the Propagation of Sheep, The Manufacture of Wool, and the Cultivation and Manufacture of Flax, with Directions for Making Several Utensils for the Business* (Williamsburg: J. Royle, 1765); also news of Wily in the *Pa. Gazette,* 5 September 1765.

54. *Va. Gazette* (Dixon), 28 September 1769, supplement.

55. Worthington C. Ford, ed., *Journals of the Continental Congress, 1774-1789,* 34 vols. (Washington, D.C.: Government Printing Office, 1904-1937), I, 75-80 (hereafter cited as *JCC*). The Association took effect on December 1, 1774. See the survey in David Ammerman, *In the Common Cause* (Charlottesville: University of Virginia Press, 1974), pp. 114-116.

56. Massachusetts in Merrill Jensen, ed., *American Colonial Documents to 1776* (London: Eyre & Spottiswood, 1955), pp. 823-825; Virginia in the *Va. Gazette* (Pinckney), 30 March 1775; and Peter Force, ed., *American Archives,* 9 vols. (Washington, D.C.: M. St. Clair Clarke and Peter Force, 1837-1853), 4th Series, II, 170-171.

57. *Va. Gazette* (Purdie), 17 February and 30 March 1775; Force, ed., *Amer. Archives,* 4th series, II, 14.

58. "A Farmer," *Address to the Inhabitants of North America* (Annapolis: Anne C. Green & Son, 1775), pp. 3-12.

59. *Pa. Gazette,* 1 February 1775; Force, ed., *Amer. Archives,* 4th series, I, 1169-1172. An earlier warning from a Berks County resident of the need for

long-term investment and development is in the *Pa. Gazette,* 14 December 1774. Also see David J. Jeremy, "British Textile Technology Transmission to the United States: The Philadelphia Region Experience, 1770-1820," *Business History Review* 47 (1973):32.

60. L. H. Butterfield, ed., *Letters of Benjamin Rush,* 2 vols. (Princeton: Princeton University Press, 1951), I, 54, 61, 74-75; also Donald J. D'Ella, "Benjamin Rush: Philosopher of the American Revolution," *American Philosophical Society Transactions,* new series, 64 (1974), part 5, pp. 43, 43n.

61. This widely reprinted speech can be found in the *Pennsylvania Evening Post,* 11 and 13 April 1775; the *Pennsylvania Magazine* 1 (October 1775):482-485; and Force, ed., *Amer. Archives,* 4th series, II, 140-143.

62. The company plan and advertisement are in the *Pa. Gazette,* 22 February, 19 April and 9 August 1775; the *Pennsylvania Packet,* 7 August 1775; Force, ed., *Amer. Archives,* 4th series, I, 1256-1257; III, 820-821. Also see the overview in Bagnall, *Textile Industries,* I, 63-72; and the list of officers in William Duane, ed., *Passages from the Remembrancer of Christopher Marshall* (Philadelphia: James Crissy, 1839), p. 16.

63. *New York Journal,* 3 March 1774.

64. *Pa. Gazette,* 16 June 1768.

65. Schlesinger, *Colonial Merchants,* p. 123; pp. 152-153, for merchants ignoring non-importation.

66. *Va. Gazette* (Purdie and Dixon), 30 June 1768.

67. Boston manufactory in the *Journal of the House of Representatives of Massachusetts* (Boston: Mass. Hist. Soc., 1973), XLIII, part 1, p. 178; glass factory in the *Pa. Gazette,* 29 August 1771 and 17 March 1773; and Bining, *Pa. Iron Manufacture,* pp. 141-142; china factory in *Pa. Gazette,* 18 November 1772.

68. *Journal of the Legislative Council of the Colony of New York, 1691-1775,* 2 vols. (Albany: Weed, Parsons & Company, 1861), II, 1530, 1540; and McIlwaine and Kennedy, eds., *Journals of Burgesses,* XI, 332.

69. These reports, dating from 1766 to 1768, can be found in the British Museum, King's MSS no. 206, a transcript on deposit at the Library of Congress (hereafter LC).

70. Albert Henry Smyth, ed., *The Writings of Benjamin Franklin,* 10 vols. (New York: Macmillan Co., 1905-1907), V, 116-117. Also see William Franklin's letter to his father of 10 May 1768 in Leonard Labaree et al., eds., *The Papers of Benjamin Franklin,* 23 vols. (New Haven: Yale University Press, 1959-), XV, 125.

71. British Museum, King's MSS no. 206, p. 7, LC Transcript.

72. Jensen, ed., *Colonial Documents,* pp. 418-420; denial by the Bostonians in the *Boston Gazette,* 9 January 1769.

73. *Documents Relative to the Colonial History of the State of New York,* 15 vols. (Albany: Weed, Parsons & Co., 1853-1887), VII, 888-889; VIII, 66.

74. The 1771 report is in PRO/CO, 5/937, fo. 62, as cited in Davies, ed., *Documents,* III (Transcripts), no. 101; the 1768 report is in the British Museum, King's MSS no. 206, pp. 38-39, LC Transcript.

75. South Carolina in Jensen, ed., *Colonial Documents,* pp. 422-423; the South in general in C. Robert Haywood, "Economic Sanctions: Use of the Threat of Manufacturing by the Southern Colonies," *Journal of Southern History* 25 (1959):207-219. See the sketch of Richard Oswald's testimony before Parliament on the

subject of American manufactures in Benjamin F. Stevens, ed., *Facsimiles of Manuscripts in European Archives Relating to America, 1773-1783*, 25 vols. (London: Chiswick Press, 1898), XXIV, 2037; also Force, ed., *Amer. Archives*, 4th series, III, 147-148; and Robert A. East, *Business Enterprise in the American Revolutionary Era* (New York: Columbia University Press, 1938), p. 20.

76. Henretta, *Evolution of Amer. Society*, p. 79; also see Bridenbaugh, *Colonial Craftsman*, p. 97; Stuart Bruchey, *The Roots of American Economic Growth* (New York: Harper and Row, 1965), pp. 26-31; and Carole Shammas, " How Self-Sufficient Was Early America?" *Journal of Interdisciplinary History* 13 (1982): 247-272, which questions the whole notion of colonial self-sufficiency.

77. *New York Journal*, 31 August 1769.

78. Thomas C. Cochran, *Business in American Life* (New York: Macmillan Co., 1972), p. 24.

2 THE ISOLATED INVENTOR

In the realm of manufacturing technology the colonies were just that—colonies. To be competitive on the levels desired by manufacturing enthusiasts they needed a constant influx of skilled workers and entrepreneurs from Great Britain and the continent, bringing with them fresh ideas, new machines, and new techniques. Those people and devices, in turn, had to be backed with capital, provided with raw materials, and set to work on a marketable product. No small task, that. Potential investors were understandably frightened away, and the home manufactures movement did not bring in a substantial profit.

Manufacturing enthusiasts were more than a little disturbed by their lack of success. As far as they were concerned, only manufactures had the potential to make the colonies self-sufficient. Self-sufficiency was their elusive economic goal, and they were willing to use any means to achieve it. They most certainly did not see their technological borrowing from the mother country as inconsistent. They concluded that borrowing would help them cultivate their own technological talents and ideally lessen if not totally eliminate their technological dependence. If those who believed in the "cement of empire" theory wanted to improve existing technologies, the radical elements who hoped that the glue would not hold were even more insistent.

Although manufacturing enthusiasts avidly sought the latest technological data from abroad, they did not do everything in their power to stimulate invention and innovation at home. They did not, probably because they had not begun to view invention as a key to technological change. Colonial entrepreneurs did not make a habit of underwriting invention because inventors did not seem to promise them much. The colonists attempted to transplant a manufacturing technology to their shores instead of build it from scratch. As a result, the role of invention in industrial progress may have escaped their notice. The many premiums offered by manufacturing societies in the 1760s and early 1770s stimulated invention in theory more than in practice. Most of the devices submitted for those premiums were copies (often imperfect) of British tools. Those tools had already proven themselves as technological innovations, so their origins and their makers were obscure. Although some of the colonies awarded patents for new devices periodically, it does not appear that the public or politicians attached much significance to the invention going on around them. Before the War of Independence, perhaps invention was too anonymous for Americans to appreciate in the way they would afterward, when the Industrial Revolution began to leave its indelible imprint on the country.

This is not to say that the colonies suffered from a lack of homegrown inventive skill. Carl Bridenbaugh has called the eighteenth century "the great age" of the colonial craftsman, and with good reason.[1] Colonial America had countless inventive artisans, many of whom designed new devices. Indeed, the challenge of early American living brought forth numerous innovations: a better axe, the long-handled scythe, and an improved rifle, to name just a few. Colonial sawyers designed new machinery, much of it water-powered, for their mills. Farmers, mechanics, and merchants tinkered in their barns, shops, and homes and turned out innumerable gadgets. They even made a few scientific instruments, such as David Rittenhouse's famous orrery. The lightning rod and stove, of course, were only two of Benjamin Franklin's inventions.

Rittenhouse did not earn his living primarily as an inventor, however, and although Franklin realized profits (which he neither sought nor kept) from his stove, the stove alone would not have earned him his reputation as the leading American scientist of the Revolutionary generation. Other inventors experienced nothing but one disappointment after another, regardless of the "utility" of their inventions. Colonial Americans, interested in invention on one level—they did, after all, buy Franklin stoves—quite possibly did not make a connection between invention and improved technologies. Artisans prospered at trades from carpentry to coopering and invented several ingenious laborsaving devices. Nevertheless, though they may have brought new tools into their trades, they did not change the nature of those trades. Artisans who tinkered in the privacy of their shops, like farmers who modified their tools, rarely made devices with larger social implications. Only occasionally did they invent tools (like the scythe in New England or devices in sawmills) that changed the nature of labor or organization of work. What modest changes they did make were only of local importance.

Specialized skills relying more on machines and factory work were just beginning to emerge in Great Britain, which was moving further into the industrial age. Information from Great Britain that would have helped the colonists tie new skills and devices with advanced technologies was withheld, poorly understood, came in piecemeal, or too close to the war to attract attention. The Board of Trade was supposed to curtail the emigration of craftsmen who could injure Britain's technological preponderance by building up indigenous, competitive colonial industries, but it did not always follow through. The Board and customs officials sloppily enforced imperial policy restricting the migration of skilled workers, the transmission of machine designs, and the seepage of new ideas out of Britain. The colonies were consequently never shut off completely from what the mother country had to offer.[2] Even so, mechanically minded colonists never had all the pieces to the industrial puzzle. As a result, adolescence in manufacturing and lack of public identification with invention as part of a more inclusive technological process extended beyond the War of Independence.

Still, the colonists were not entirely ignorant of the factors contributing to technological progress even if they had much to learn about the details of large-scale technological change. The same voices calling for home manufactures recognized the need for technological transfer and pressed for the importation of skilled workers. Echoing Thomas Barnard, "Phileleutheros" counseled that the solution to home manufacturing problems lay in recruiting British workers because they possessed a technological expertise unavailable in the colonies. He admitted that the initial cost of attracting them would be high, but he was confident that the profits to be made would more than return the investment.[3] Phileleutheros and others like him welcomed emigrant craftsmen with open arms and gushing promises of success.

European artisans of all kinds—flaxworkers, spinners, weavers, ironworkers, sawyers, miners, and masons—sought transatlantic passage, despite imperial strictures against it. They usually made a point of letting the public know their origins. A Yorktown, Virginia, brass and ironworker advertised that he had "Materials and Men, from the best shops in London."[4] An Irish linen expert setting up shop in Philadelphia noted with pride his ability to dress flax and hemp "according to the most approved method" used in Europe. Another arrival in that same town volunteered to build and operate a woolen manufactory if investors would put up the capital.[5] A London wool card maker settling in Boston advertised that he made "all the different Kinds of Cards made in Great Britain," while another Englishman offered his services to New Yorkers as manager of a textiles factory.[6] Bostonians negotiated with several English fullers and dyers to help with their factory in 1765.[7] And New Yorkers engaged "thirteen of the best Hammer-men and Forge-men" in Sheffield, with the guarantee of high wages for at least two years, a cash bounty, and an allowance for those leaving their families behind.[8]

The foregoing accounted for but a trickle of the flood of inventive craftsmen pouring into the colonies in the 1760s and early 1770s. Many made the crossing as indentured servants. In May 1766 a ship from

Dublin docked at Philadelphia with a passenger list of indentured servants that included coopers, tanners, carpenters, tailors, weavers, braziers, and a wool comber. Three years later the *Pennsylvania Gazette* repeated claims that "workmen and manufacturers" of every kind arrived continually at New York and Philadelphia; so many, in fact, that "in a few years they shall be able to supply every necessary among themselves."[9] Rumors also circulated that an English linen manufacturer planned to bring over six journeymen and all his machinery to the colonies. One New Yorker smuggled in machinery for rolling and "refining" copper, lead, and iron. Encouraged by members of the House of Burgesses, Elisha and Robert White erected a "Woollen and Worsted Manufactory" in Hanover County, Virginia, in 1774. To carry off their enterprise they had "at a great expense, sent to Great Britain for a number of the best workmen."[10]

Nonetheless, flood or no flood, the colonies could never bring over all the inventive craftsmen needed to build up industry to the levels desired by manufacturing enthusiasts of the 1760s and early 1770s. To make matters worse, some emigrant craftsmen left their trades to farm on the frontier, a movement Thomas Barnard warned of and royal governors gloated about. The frontier, while stimulating invention in some areas, crippled it in others because of this.

Frontierless England had just the opposite problem. There some pre-Luddite mobs had already wrecked machines in their fear of being displaced.[11] Americans, not identifying with the motives behind those acts and largely unfamiliar with the beginnings of industrial blight in Manchester, had no such fears. Franklin and Rush, who came closest to opposing large-scale manufacturing, were at best ambivalent on the subject. They wrote enthusiastically about home manufactures one moment and related their misgivings about factory work the next.[12] Although they objected to following the British model too closely, if following that model meant the rise of an urban laboring class, they nonetheless wanted to pick and choose those machines and techniques that would make American manufacturers more competitive. Their misgivings were hardly noticed because industrialization proved to be attractive in a land both loved and feared for its vastness. Industrialization became just that much more desirable to a people who felt like tiny cogs in the vast machinery of empire. The lure of western lands may have proved deleterious to the quick advance of colonial manufacturing, but in the long run it proved beneficial by aggravating the desire for laborsaving machinery.

Ironically enough, gentlemen farmers and merchants in the mother country, without realizing the long range implications, sometimes helped the colonists in their quest for technological information. A Liverpool resident sent instructions on processing hemp to the House of Burgesses. The New York manufacturing society organized in 1764 received donations and "useful Hints" from several Londoners.[13] Some colonists, not content with secondhand information, went to England to learn for themselves. Planter-merchant James Stewart of Winchester, Virginia, a self-styled champion of home manufacturing, traveled around England for eighteen months. He gathered data on dyes for linen, cotton, and wool cloth and specimens of roots and shrubs used for the dyes. Upon returning home he sold some plants and instructed tradesmen in the construction of a new spinning machine.[14]

Colonial newspapers kept their readers posted on the activities of the Royal Society for the Encouragement of Arts, Manufactures and Commerce. Founded in 1754, the Royal Society was one of the first to encourage inventors and award premiums for a wide variety of mechanical devices, from agricultural machinery to diving bells. It also awarded premiums for improvements in agricultural and industrial processes. No less important, it served as a model for the manufacturing societies that sprang up in the colonies in the 1760s.[15] Colonial newspapers and magazines followed other British scientific societies and from time to time published extracts from their journals.[16]

Dozens of colonists sought direct ties with the British societies. By 1772 over forty, Franklin included, were corresponding members of the Royal Society. Franklin kept in touch with the society, as did others who sometimes sent along news of inventions made by fellow colonists.[17] The New York manufacturing society founded in 1764 opened a correspondence, hoping to stimulate a transatlantic exchange of information. Although the Royal Society had limited resources, it encouraged such hopes and went so far as to grant Georgia over £1300 for silkworm-raising experiments.[18]

The colonial organization that did most to promote the transit of technology and generate interest in invention was the American Philosophical Society, formed in Philadelphia in January 1769. The origins of the society extended back to 1727 and Benjamin Franklin's secret club the "Junto." The Junto broadened its membership in 1743 and finally, after languishing two decades, evolved into the leading scientific society in early America.[19] The American Philosophical Society was built around Franklin, the colonial scientist par excellence, and a talented group of devotees that included David Rittenhouse, Owen Biddle, and Dr. Thomas Bond.

The society had as its purpose the promotion of science in general and practical knowledge in particular, those "Subjects as tend to the Improvement of [this] Country, and Advancement of its Interest and Prosperity." Learned men, society members believed, should take the lead in introducing new ideas and techniques. "The Bulk of Mankind follow a beaten Track," they chided; men "seldom turn their Thoughts to Experiments, and scarcely ever adopt a new Measure" unless certain of success. The society encouraged all colonists—farmers, mechanics, and professional men—to experiment with progressive farming techniques, develop new plant strains, and invent new mechanical devices for agriculture and manufactures.[20] The American Philosophical Society assigned an important role to invention in bringing progress. It served a useful function by publicizing the work of "ingenious artists, who might otherwise remain in obscurity." The society, in the words of member William Smith, encouraged "such Mechanic Arts, Inventions and useful Improvements, as tend to shorten labor, to multiply the Conveniences of life, and inrich the community."[21]

Promotion of invention had for years been a pet project of Franklin, the society's first president. His plan for the Academy of Philadelphia, which opened its doors in 1751, stated that young men needed to study mechanics, "that art by which weak men perform such Wonders, labour is sav'd, [and] manufactures expedited." His circular letter of May 1743 calling for a philosophical society had likewise urged the study of "New Mechanical Inventions for saving Labour; as Mills, Carriages, &c., and for Raising and Conveying Water," in effect, all mechanical improvements that "tend to increase the Power of Man over Matter, and multiply the Conveniences or Pleasures of Life."[22]

The society proposed an ambitious program of regular correspondence with the Royal Society of Arts, publication of its own transactions, and communication with interested groups in the colonies. Of the six original standing committees in the American Philosophical Society, two dealt directly with invention: the committee on mechanics and architecture, and the committee on husbandry and American improvements.

Because of the society, Philadelphians saw themselves as residing at the seat of American science, and with some justification. Francis Hopkinson's eulogistic lines proclaiming "Fair Science, softening, with reforming Hand, the native Rudeness of a barb'rous Land" were composed with Philadelphia in mind.[23] One Philadelphian exclaimed that no city could rise faster than by "excelling in *Works of Genius and Useful Arts*." In his opinion, "wise Legislators" should back them to insure continued progress. The Pennsylvania Assembly had led the way by financing the observations of the transit of Venus and awarding Rittenhouse £300 for his orrery.[24] Members of the assembly consistently encouraged the philosophical society. So did other politicians, notably the Philadelphia county commissioners, who presented it with a model of a machine for weighing loaded wagons.[25] Philadelphians were kept posted of the society's activities in Franklin's old paper, the *Pennsylvania Gazette*. It is no coincidence that notices on inventors first began appearing in print soon after the society's founding.

The philosophical society, then, helped laymen take note of invention. Science was made a community project, and scientific achievement became a source of public pride.[26] Undoubtedly many Pennsylvanians shared Thomas Paine's view that the American Philosophical Society, "by having public spirit for its support, and public good for its object, is a treasure we ought to glory in."[27]

Society members took part enthusiastically in the home manufactures movement. In 1770 they suggested to the Pennsylvania Assembly that the introduction of the silk culture would bolster textile manufactures and, as a result, the provincial economy as a whole. With that body's support, prominent society members joined with a few others to raise capital for their proposed silk company. They secured mulberry trees for the cocoons from Italy, exchanged seeds and ideas with a silk-growing society in Connecticut, and offered premiums. Ultimately the men behind the scheme, such as Owen Biddle, hoped to export silk to Britain. They enjoyed some early success and paid premiums through 1774. Nevertheless, though fine quality silk was produced, it was produced in small quantities.[28] The war brought an abrupt end to the silk company, and the society eventually sold its assets.

Society members threw themselves into other projects as well. They tried, without avail, to keep the faltering Philadelphia paper factory from closing down in 1773.[29] They promoted two canal projects to improve inland navigation: one to link the Delaware River with Chesapeake Bay, the other to connect the Susquehannah and Schuylkill rivers.[30] The war interrupted work on both canals, which had hardly progressed beyond the planning stages.

By July 1773 Virginia had a scientific society with a practical bent paralleling that of the American Philosophical Society. This new association grew out of an organization founded in Williamsburg in 1759 to promote manufacturing with premiums and bounties. The "Virginia Society for Advancing Useful Knowledge," as it became known, was pledged to the "Study of Nature, with a View of multiplying the Advantages that may result from this Source of Improvement."[31] The *Virginia Gazette* published a note from one "ACADEMICUS" expressing relief that such a society had finally been formed. Perhaps, this writer speculated optimistically, it would bring some needed advances in mechanics and agriculture and make the colonies bona fide members of the international scientific community. He lamented that heretofore "there had not been a single Production which has occasioned our Name even to be known in the Republic of Letters."[32]

The castigation of "ACADEMICUS" can be viewed in one of two ways. From one angle his judgment could be dismissed as petulant. After all, Franklin was respected as one of the virtuosi, a pioneer in basic science with his work on electricity and an inventor of many useful devices. Linneaus himself called John Bartram a great botanist, and David Rittenhouse drew flattering comments from many circles on both sides of the Atlantic. Americans in general had a reputation for pragmatic ingenuity, for a utilitarian approach to science that assisted them in carving a civilization from the wilderness. Yet when viewed from a different angle, "ACADEMICUS" had a point. No colonial inventor had designed a new machine capable of revolutionizing manufacturing or fundamentally changing the colonial economy. To that extent he expressed the frustrations of those impatient to do more. This impatience became more pronounced as Americans began considering political independence.

The Crown and Parliament at least partially realized the potential danger of allowing the colonies to become too technologically independent. From 1718 on, Parliament passed legislation curbing the emigration of textile workers and other artisans. In 1774 textile machinery was added to the list. Some patriots recognized the motive behind this legislation and cried out in alarm.[33] Still, as already noted, Parliament and the Board of Trade enforced these acts rather irregularly. Possibly British statesmen prior to the War of Independence were themselves only vaguely aware of the industrial transformation going on in their midst.

Although many colonists linked technological with political independence, they were unsure how invention could help them reach that goal.[34] True, they appreciated the value of new tools and took pride in innovations such as the Pennsylvania rifle and Conestoga wagon. But they did not see invention as part of a technological process, a way of stimulating fundamental change. Inventions that could revolutionize manufacturing, such as Richard Arkwright's water frame, or alter the very basis of work and culture, such as James Watt's new reciprocating steam engine, remained mysteries to most colonial Americans. The War of Independence came just when entrepreneurs in England like Matthew Boulton were drawing together the most important inventions of several decades. Americans had to wait until after the war to learn about the kind of manufacturing made possible by steam-driven machinery.

Before the war Americans kept abreast of new inventions, but inventions of a character far different from Watt's steam engine or Arkwright's water frame. Up until the 1770s most colonial newspapers gave more space to reports from Europe than to news from around the colonies. Included in those reports were short articles on mechanical devices, devices qualifying more as "curiosities" than inventions of real social import. Of course there were exceptions. The papers did carry information on the canal-building boom in England. They reported useful inventions like a wind-operated ship's pump, an improved spinning wheel in Ireland, desalinization experiments along the English Coast, and sundry hydrometers, windlasses, dyes, and advances in ship construction.[35] These reports were nevertheless outnumbered by and mixed in with articles on inventions ranging from the curious to the bizarre. They included musical chairs, itch cures, and chemical concoctions, with one reported to make leather hard as armor, and another to make steel soft enough to be sliced with a knife.[36] There were accounts of a cannon made of "animal jelly and tow" and of another piece of ordnance capable of firing fifty balls at once.[37] These were joined by stories about a Swedish rocket boat and a leather and wood French submarine that purportedly navigated underwater in the Bay of Biscay for over four hours.[38]

American magazines and newspapers, as well as the correspondence of leading men of science, made no reference to James Hargreaves, Watt, or Arkwright. Even Franklin, a correspondent of Matthew Boulton's, did not allude to the role of invention and Britain's patent system in fostering the industrial Revolution, nor did he do much to transfer steam technology to the colonies. Colonial magazines and papers reprinted in detail news about European diplomatic and military affairs but virtually nothing about the industrial transformation of the mother country. This resulted from the secrecy of some inventors, imperial policy, and the fact that Britain was beginning to industrialize just a few years before the war, when such news was deemed less pressing than reports on the political crisis within the empire. Colonists thus knew very little about the beginning of the Industrial Revolution and the central role invention played in it. Ignorant of the setting of which those inventions were a part, the colonists treated them as curiosities when they did learn of them. As Carroll Pursell has pointed out, interest in steam engines, for example, "was limited largely to those who had a philosophical turn of mind and the leisure and opportunity to dabble in science." Before the Revolution only three steam engines were used in the colonies, and they were Newcomen, not Watt, models. All three enjoyed undistinguished careers, in part because Colonial America "provided neither the deep mines nor the large urban waterworks which would have made steam engines economically attractive."[39]

Americans lacked a frame of reference, a technological context into which they could fit important inventions after hearing of them. Technologically, Watt steam engines, like Arkwright frames, were significant only after being worked into a form of factory production new to Britain and totally lacking in the colonies. And it was only with the Industrial Revolution, when the new machines—the inventions—of Watt and the rest were integrated into a mode of production by entrepreneurs like Boulton and Arkwright, that invention began to play a visible role in technological change.

Not in touch with inventive processes in Britain that promoted industrialization, the colonists were similarly out of touch with the process of invention on their own side of the Atlantic. A New Englander calling himself "A.Z." attempted to alert his fellows to the need for a serious appreciation of invention. He drew a direct connection between invention and prosperity and drove his point home in an allegorical "genealogy of commerce." According to this allegory, "Necessity" was a "female bastard of an ancient family" begotten by Pride, the father, and Sloth, the mother. She married Poverty, who fortunately was "a likely fellow," and together they sired Invention, a son, and Witt, a daughter. Because of his genius, Invention cared for the entire family and in turn produced a number of his own children, notably, Industry and Ingenuity. Ingenuity begat Barter, and Barter begat Trade. Trade, the commonly acknowledged key to colonial prosperity, was made the figurative great grandson of Invention.[40]

The meaning of this allegory may have escaped a society not accustomed to assigning invention a place in technological growth. It was not that most Americans were uninterested in invention, it was that they did not see invention in the same light as A.Z. An early magazine, *The American Magazine and Historical Chronicle,* first published in 1743, had promised to keep its readers informed of the latest scientific and "mechanical improvements" in Europe and the colonies. Yet the only mechanical device reported during the magazine's three years of publication was an orrery donated to Yale College. The Philadelphia based *American Magazine and Monthly Chronicle,* appearing from 1757 to 1758, had a section on "Philosophical Miscellany" devoted to articles or news about science and the "mechanical arts." Of the twenty-five pieces in this section all but two dealt with comets, earthquakes, electrical experiments, eclipses, and the like, usually reprinted verbatim from European journals. Of the two exceptions, one reported the London society's premiums for 1758, and the other discussed Thomas Godfrey's improvement of the quadrant. Godfrey was the only colonial inventor to draw notice; the magazine made no effort to dig up information on others. Lewis Nicola's *The American Magazine,* though lasting less than one year, gave considerable space to scientific subjects ranging from notes on the American Philosophical Society to reprint articles on flaxdressing and raising silkworms. Yet even Nicola's magazine had nothing on invention.

When the *Pennsylvania Gazette* compiled a list of "American Inventions" in 1771, it consisted of four items: Franklin's lightning rods, Jared Eliot's essay on black sand, a theory on comets developed by John Winthrop of Harvard and a mercurial inoculation developed by a Long Island doctor.[41] Nowhere is there mentioned the Pennsylvania rifle, the broadheaded axe, the long-handled scythe, or the Conestoga wagon. Not only was the list short, it reflected a loose application of the word "invention."

The list did not do justice to American ingenuity, but it did indicate prevailing thought or, better yet, lack of thought about the social significance of invention. Colonial society did not regularly promote invention because invention did not seem to be especially important to colonial prosperity. Support of invention was haphazard, and that haphazardness was reflected in colonial patent granting practices. This does not mean that patents are the source of invention or even the most important stimulus to technological innovation. Nevertheless, patents do reflect a certain awareness of the importance of invention, whether or not they hurt individual inventors while helping others.[42] Patents show the social worth attached to the *process* of invention. Governments instituting patent codes make invention part of the public domain even as they protect the property of individuals. By granting protection and privileges to invention as a form of "intellectual property," patents give inventors special legal status. They are given that special status because of their recognized value in designing laborsaving machines, machines that they could promote as agents of social welfare.

In the seventeenth century Great Britain had taken the lead in issuing patents to inventors. Francis Bacon urged the Crown to grant them in 1602, and in 1623 James I issued the first fourteen-year patents. The Statute of Monopolies passed by Parliament in 1624 marked the full emergence of the patented invention, which heretofore had been mixed in with grants of royal favor for industrial privileges or import monopolies. By the end of the seventeenth century, patenting inventions had become common practice and by the middle of the eighteenth century, the patent system had matured enough to play a part, however small, in promoting the Industrial Revolution.

Patents or "monopolies" in the colonies were much rarer, with some colonies making provisions for them and others not.[43] The Massachusetts "General Lawes" of 1641 had a patent clause stating, "There shall be no *Monopolies* granted or allowed amongst us, but of such new inventions that are Profitable to the Country, and that for a short time." Connecticut followed in 1672 with a similarly worded statute. By 1691 South Carolina had a patent clause "for the better encouragement of the making of engines" for farming.[44]

Joseph Jenkes had the distinction of receiving the earliest colonial patent for a mechanical device. Jenkes, a versatile master craftsman, was granted a fourteen-year monopoly on all water-powered mills by the Massachusetts General Court on May 6, 1646. He went on to design a scythe, which he patented in 1655, and to build a fire engine and sawmill.[45] His work, however, did not trigger an avalanche of requests for, or bequests of, patents. The prolific Jenkes was the exception proving the rule, for patents were not at all common in Colonial America. Patents were not treated as part of a larger technological process, as they were in Britain. There are only scattered references to "patents" or "monopolies" in colonial records. In 1728 Samuel Higley and Joseph Dewey of Connecticut received a ten-year monopoly on a steelmaking process they had discovered. In 1753 two other Connecticut residents, Jabez Hamlin and Elihu Chauncey, secured a patent for dressing flax with water-driven machinery, machinery copied from models used in Scotland and Ireland. In 1774 John Shipman patented a flour-grinding tidal mill.[46]

There were other cases where inventors secured some sort of protection, but patents for invention were on the whole uncommon. The type of patent granted by Connecticut to Hamlin and Chauncey, though somewhat more common, was designed as an industrial privilege, a monopoly, rather than as a reward for originality. Since monopolies carried an onus of "special privilege" that galled many colonists, even they were granted carefully. Pennsylvania passed a resolution in 1717 officially recognizing the transfer of patent rights of a Pennsylvanian granted in the mother country, but it did not grant patents of its own until the 1780s. A clause in the 1682 Frame of Government for Pennsylvania promising to "encourage and reward the authors of useful and laudable inventions" was never employed.[47]

Although the South Carolina legislature allowed its 1691 patent statute to lapse by 1736, it did issue patents now and then. In 1743 it passed an act "for the encouragement of Mr. John Timmons in his projection of a new instrument for cleaning of rice." In 1755 and again the next year the legislature granted patents for new rice machines. Since the legislature also passed laws encouraging rice growing, it is fairly clear that at least some South Carolinians recognized the close tie of invention to agricultural progress.[48] But apparently the only inventions receiving any recognition or protection were those that could be used to improve the rice harvest. Rather than promote invention to introduce new crops or products, the legislature did so to stimulate a business already in operation. Invention had to fit into an existing form of enterprise before receiving encouragement. Patents were issued as special legislation; there was no standardized, uniform code.

Colonial inventors, instead of seeking patents, more often applied for premiums or "awards" which usually took the form of a small cash prize, granted in one lump sum. Colonial legislatures periodically passed acts to stimulate manufactures or agriculture, and inventors attempted to have their genius recognized by demonstrating the immediate financial practicality of their ideas. Some managed to garner contracts for special projects, such as Hans Christiansen of Bethlehem, Pennsylvania, who filled a contract to build a municipal waterworks by making wood pipes and self-acting pumps to carry water from a holding tank in the town square.[49] In 1774 Christopher Tully assembled what was reportedly the first cotton and wool spinning machine made in the colonies (although it was an imitation of a Hargreaves jenny). It was used in the factory of the United Company of Philadelphia. Tully and John Hague, an immigrant weaver from Derbyshire, were awarded £15 apiece by the Pennsylvania General Assembly for their jennies, both of which had been constructed from memory.[50]

Pennsylvania had more than its share of inventors who, like Tully and Hague, tried to cash in on their work. More than a few approached the American Philosophical Society. They hoped to gain approval there before petitioning the provincial government for recognition or money. Between 1769 and 1775 the society accepted for examination a flax dressing machine, a model of a new windmill, a fireplace, a bridge model, and a horsedrawn multiple scythe machine.[51] It did no more than profess an opinion on the practicality or impracticality of a project, so inventors learned not to read too much into its pledge to promote invention. The society served as a valuable forum for new ideas and devices, but unlike the Royal Society of Arts, it did not offer financial rewards or the quasi-official recognition that the inventors desired. Consequently, some inventors applied directly to the Pennsylvania General Assembly. Few found remuneration there either.

Arthur Donaldson was one of a handful to gain recognition from the American Philosophical Society and a stipend from the Pennsylvania Assembly. He invented a clamshell dredge called the "Hippopotamus," an oval-shaped, flat-bottomed boat approximately thirty feet long by twenty wide. It had a scoop supported by beams and guided by ropes and pulleys, all of which were powered by a horse and activated and released by pins and levers. The horse moved in a twenty-foot circle around a capstan, to which ropes holding the scoop were fastened and wound by the horse's movement. The dredge dug up river mud and dropped it into a scow anchored alongside.[52]

Donaldson notified the American Philosophical Society of his work in March 1773, when he presented it with a scale model of the dredge. The society, "desirous to encourage [so] usefull an Invention," appointed a committee to study the "Hippopotamus." Over the space of two years, three different committees examined the model.[53] They apparently delayed passing judgment until Donaldson could make a full-sized boat. When completed, the dredge not only impressed David Rittenhouse, Owen Biddle, and other committee members, but the normally tight-fisted General Assembly awarded the inventor £100 for "his Ingenuity."[54] This recognition, limited though it might have been, was far above what most inventors could hope to receive.

A few Virginia inventors also succeeded in securing official notice. In 1764 Aaron Miller asked the House of Burgesses for a "reward" for his surveyor's compass and protractor. A committee chaired by Richard Henry Lee decided "that the Invention is ingenious and deserves the Publick's Encouragement" and recommended that Miller be given £30.[55] John Hobday of Gloucester County invented a "cheap and simple Ma-

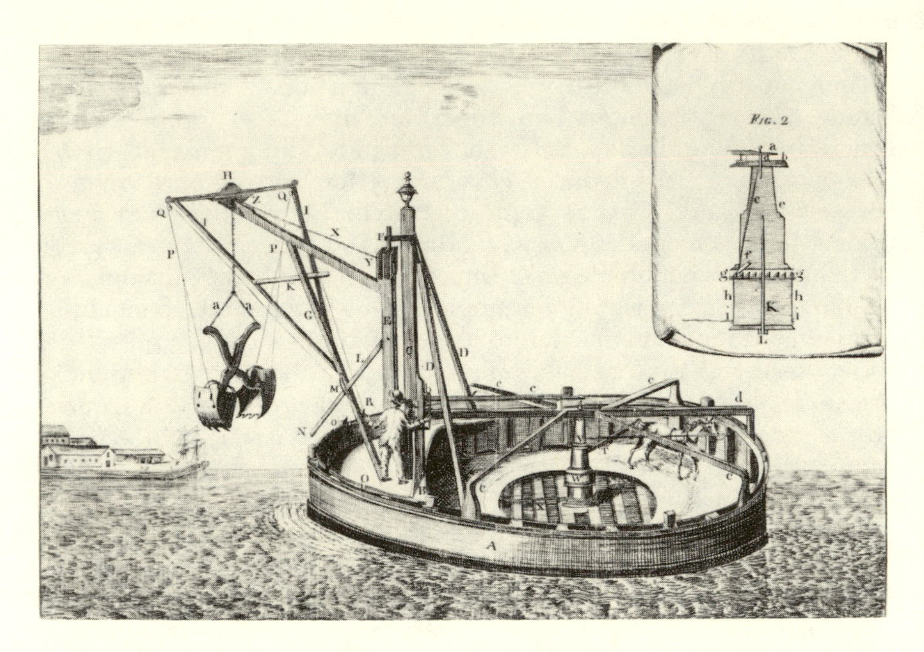

3. Arthur Donaldson's river dredge, the "Hippopotamus," as engraved for the *Pennsylvania Magazine,* May 1775. Courtesy of the William L. Clements Library, University of Michigan.

chine for separating Wheat from straw" which was capable of threshing 120 bushels a day. Far from keeping the details of construction a secret, Hobday and his brother toured several tidewater counties in 1772, giving instructions on how to make the machine. All Hobday sought in return was an initial fee from planters intending to use his device.[56] In 1774 he petitioned the House of Burgesses for a reward, as had Miller. A committee investigating his claim recommended that he receive £300, and the sum was reduced to £100 after a sharp floor debate. Shortly thereafter the Virginia Society for the Advancement of Useful Knowledge voted Hobday a medal and a "pecuniary Reward" for his "very ingenious and useful Machine for threshing out Wheat."[57]

Other Virginia inventors chose not to apply to the House of Burgesses or the scientific society. George Walker of Brunswick County invented a threshing mill for wheat, barley, oats, and rye in 1770. So did William Massie of New Kent County two years later. Rather than seek official support, they tried to finance their work by gathering "subscriptions" from wealthy planters. It does not appear that either inventor sparked enough interest to make production profitable.[58] Indeed, even Hobday had difficulty interesting his neighbors in adopting his new machine. All three were disappointed by the reception they met. Instead of being eager to use the latest improvements, most planters remained content to do things as they always had.

Countless inventors never sought official assistance through premiums or protection through patents. Most colonial assemblies did not liberally dispense either one. Pennsylvania resident John Sellers, a member of the American Philosophical Society, invented a "rolling screen" for flaxseed. Having no recourse to a protective patent, he licensed individuals to sell his rolling screens for him. It did not take long for others to copy the screen and destroy his short-lived monopoly.[59] Richard Wells of Burlington, New Jersey, another inventor and member of the American Philosophical Society, designed a special release spring for carriages. He made no attempt to advertise his innovation or capitalize on it in any way.[60] The next year he invented a wind-operated ship's pump, a model of which he deposited with the philosophical society.[61] He had numerous other ideas and innovations he communicated only to members of the philosophical society.[62]

Wells and other colonial inventors found it difficult to sell their new devices. The career of Christopher Colles best typifies the frustrations of those who longed for something more promising. Indeed, the ingenious Colles "was one of those hapless individuals upon whom fate plays unkind tricks."[63] A native of Dublin, Ireland, Colles crossed over to the colonies and settled in Philadelphia. He tried to capitalize on knowledge gained from a good education as a youth and engineering work on a canal in Ireland. Calling himself an "Engineer and Architect," he advertised his willingness to construct "mills of various kinds, and other Hydraulic Engines." He also offered "to instruct young Gentlemen at their Homes in the different Branches of the Mathematics and Natural Philosophy."[64]

For two years Colles scrimped for a living as a private tutor and public lecturer. His application to the Pennsylvania General Assembly "praying Assistance . . . in providing himself with an apparatus for a Course of Lectures on the Science of Mechanics" was tabled without comment.[65] Undeterred, the Irishman sought to interest the public in his inventions. He launched a newspaper advertising campaign:

> IRON FURNACES. In all new established Colonies it appears highly necessary to use every means for moderating the price of labour, and for carrying on all kinds of manufactures with the greater facility, to which purpose mechanical contrivances seem particularly adapted; and as CHRISTOPHER COLLES, of Philadelphia, has made that branch of knowledge his favourite study, he gives the public notice that he has lately thought of some things, and made some new inventions, which he apprehends may be exceedingly advantageous to this country.[66]

Colles went on to describe two inventions, one a furnace for extracting

4. Christopher Colles, technological visionary. Courtesy of the Historical Society of Pennsylvania.

ore, the other a "cheap and simple machine to raise water." He had already contracted with a Philadelphia distillery to build and install a steam engine for raising water. In August 1773 he invited members of the American Philosophical Society to inspect his handiwork. A four-man committee, which included inventors David Rittenhouse and Richard Wells, went to see Colles's Newcomen-style engine in action. They found that, owing to flimsy materials, the engine was unsound. But they also admitted that the engine's unsuitability was due to Colles's lack of money, not expertise. They reported back that "the undertaker is well acquainted with the principle of this particular branch of Mechanics" and "therefore worthy of Public encouragement."[67] Nothing further happened. The society had no funds for a new engine, and neither did Colles. The city fathers of Philadelphia apparently could not be interested in the project. Philadelphia may have been the seat of American science, but Philadelphians did not automatically throw their support behind every new project. In Colles's case this was unfortunate, since Colles recognized quite clearly the potential of laborsaving devices in the colonies.

Three years of work in Philadelphia had brought Colles a pittance, barely enough for his family to live on. They moved to New York City where Colles persuaded the New York Common Council to employ him as engineer on a mammoth municipal waterworks project. In 1774 the council granted Colles £2600, half of which went for a Newcomen-type steam engine, the other half of which went for a reservoir and pipes. The engine pumped water from a pond up to a hillside reservoir. From there the water coursed through fourteen miles of hollowed-out pine pipes buried beneath the streets of New York. The system was completed by April 1776 and was working, albeit rather imperfectly, when the British occupied the city in September. It fell into disrepair soon after, and Colles, possibly on the verge of striking success, fled the city to become enmeshed in the war.[68]

No other colonial inventor came so close to making a name for himself on a project of such dimensions. If the waterworks had operated successfully for a while longer, with the kinks and defects gradually eliminated, it might have demonstrated to some the potential of steam power and, just as important, the role of invention in mechanization and technological change. This, however, was not to be.

Colles encountered the full range of responses to invention in colonial society. He, like dozens of other inventors, found a receptive audience for some projects, though not for others. He also found that scientific curiosity alone could not guarantee success, nor could endorsement by his fellow inventors. His work in a Philadelphia distillery showed the need for capital to pay for new devices. His New York waterworks project

As the ſeveral Inhabitants of this City are particularly intereſted in the following Affair, it is therefore judged proper to lay the ſame before them.

COPY OF A

PROPOSAL

OF CHRISTOPHER COLLES,

For furniſhing the City of *New-York* with a conſtant Supply of FRESH WATER.

To the WORSHIPFUL

The MAYOR, ALDERMEN, and COMMONALTY,

Of the City of NEW-YORK, in COMMON COUNCIL convened.

THE numerous and important Advantages which great and populous Cities derive from a plentiful Supply of freſh Water, requires a general Attention; and as this City is very deficient in this Article,

CHRISTOPHER COLLES,

HUMBLY offers his Services to erect a Reſervoir on the open Ground near the New Gaol, of One Hundred and Twenty-ſix Feet Square, with a good Bank of Earth ſurrounded with a good Brick or Stone Wall Twelve Feet high, and capable of holding One Million Two Hundred Thouſand Gallons of Water; which will be of exceeding Utility in Caſe of Fire, which all Cities are liable to. To erect a Fire-Engine in a good Brick or Stone Houſe cover'd with Tiles, capable of raiſing into the ſaid Reſervoir Two Hundred Thouſand Gallons of Water in Twenty-four Hours. To lay Four Feet deep through the Broad-Way, Broad-Street, Naſſau-Street, William-Street, Smith-Street, Queen-Street, and Hanover-Square, a main Pipe of good Pitch Pine of ſix Inches Bore, well hooped at one End with Iron; and through every other Street, Lane and Alley in the City South Weſt of Murray's-Street, King George's-ſtreet, Banker's-Street, and Rutger's-Street, the like Kind of Pipe of Three Inches Bore, with a perpendicular Pipe and a Cock at every Hundred Yards of ſaid Pipes,---a proper Contrivance to prevent the ſame from Damage by Froſt; and alſo on every Wharf a convenient Pipe and Cock to ſupply the Shipping. The Whole to be completely finiſhed in a workmanlike Manner within two Years from the Time of making the Agreement, for the Sum of Eighteen Thouſand Pounds New-York Currency, by

CHRISTOPHER COLLES.

The following Calculation ſhewing the Utility of the above Deſign, will, it is imagined, be found upon Inſpection as fair and accurate as the Nature of ſuch Things will admit.

It is ſuppoſed there are 3000 Houſes that receive Water from the Tea Water Men; that at the leaſt, upon an Average, each Houſe pays One Penny Half-penny per Day for this Water; this makes the Sum of £. 6750 per Annum, which is 45s. for each Houſe per Ann. According to the Deſign propoſed, there will be paid £. 6000 per Ann. for four Years, which is 40s. each Houſe: By which it appears, that even whilſt the Works are paying for, there will be a ſaving made to the City of £. 750 per Ann. and after the ſaid 4 Years, as the Tax will not be more than 10s. per Annum to be paid by each Houſe, it is evident that there will be ſaved to the City the yearly Sum of £. 5250, for ever.

In this Calculation it is ſuppoſed that 40s. per Ann. is to be paid for 4 Years, but this is done only to provide againſt any unforeſeen difficulties that may occur. It is imagined that that Sum paid 3 Years will effect the Buſineſs. The great Plenty of the Water, and its ſuperior Quality, are Advantages which have not been before ſpecified, but muſt appear of conſiderable Moment to every judicious Perſon.

NEW-YORK: Printed by HUGH GAINE, in HANOVER-SQUARE.

5. Christopher Colles's New York water supply proposal, published in 1774 after the plan was accepted by the city council. Courtesy of the Chapin Library, Williams College.

did the same and more. There he needed time to modify and improve his devices as well as money to pay for failed experiments.

The colonies had no shortage of inventive geniuses. Colles was one of several to experiment with steam engines; John Hobday was one of many to experiment with new farm machinery; Richard Wells was but one of hundreds of inveterate tinkerers. Collectively they were proof that "Yankee ingenuity" was a vital part of the colonial experience. But the ingenious Yankees did not work a technological transformation of colonial society. Those hoping to introduce significant technological change—like Colles—were caught in a society in flux. From an inventor's standpoint, the colonies had compiled an uneven record. Although colonial living stimulated the genius of many and brought forth a variety of new and useful tools, colonial society had not always nourished that genius or adopted those tools. One Pennsylvanian lamented "that the mechanic Arts and Manufactures cannot be encouraged by our Legislature with the same Propriety that they promote the liberal Arts and Sciences."[69] For this he blamed Britain and the imperial connection circumscribing American political and economic life. Benjamin Franklin lashed out more vehemently at what he saw as the shabby treatment accorded inventors. "There are everywhere a number of people," he complained, "who being totally destitute of inventive faculty themselves, do not readily conceive that others may possess it." Regrettably those people had their sentiments confirmed "by frequent instances of pretensions to invention, which vanity is daily producing."[70]

The American Philosophical Society aided some inventors, but it had little money and was in a province where patents were not granted. Colonial assemblies sometimes lent assistance, but not consistently. Given the youthful immaturity of colonial society and the disadvantages attending that youthfulness, the scientific societies and assemblies who took the initiative to support or recognize inventors did remarkably well. They had made a significant beginning. John Hobday did not earn much money from his new farm machinery, but he did receive recognition from the Virginia Society for the Advancement of Useful Knowledge and a stipend from the House of Burgesses. Christopher Colles did not enjoy the success he had expected, but he did stir up some curiosity in Pennsylvania and New York. If the Pennsylvania General Assembly had snubbed Colles, the New York City Council had more than compensated by giving Colles its utmost confidence.

A few individuals did what they could to help inventors. As editor of the *Pennsylvania Magazine*, Thomas Paine saw to it that news of inventions was printed. He and the contributors to the magazine were, for the most part, zealous patriots. They included notes on inventions because they wanted to boost national pride. They wanted their readers to develop a national identity, and promoting invention was one way of doing it.

From the first issue in January 1775 to the last in July 1776, the *Pennsylvania Magazine* strove to plant the seed of nationalism. It was only fitting that the Declaration of Independence was reprinted in the last issue. Paine's ideas about invention and the promotion of American nationalism were of a piece with his sentiments in *Common Sense*, although Paine probably did not see all the technological implications of his arguments. Publicly, at least, he made no connection between invention and the quest for technological independence undertaken by the manufacturing enthusiasts.

The American Philosophical Society, like the Royal Society of Arts it was patterned after, promoted both invention and manufactures. It was not far from pairing the two. The American Philosophical Society backed science and technology for the good of "all mankind" and for Americans in particular. It espoused a belief in the international nature of science, yet it also espoused a nationalism not all that different from that of the owners of the *Pennsylvania Magazine*. Its commitment to promoting the work of "ingenious artists, who might otherwise remain in obscurity" led the way for a nation that eventually came to share that commitment, but that national interest in invention was still a decade away.

Fascination with invention was restricted to a comparatively small group of scientifically minded colonists. Most colonists did not take note of new inventions because inventors were not generally seen as essential agents of technological change. Besides, Americans were only just beginning to tie technological change to their visions of social progress. Inventors by and large worked in isolation. The innovations of countless gunsmiths, millers, and mechanics, which undoubtedly recurred constantly, were too subtle, too anonymous, to strike the public fancy. Franklin's observation on the plight of inventors perhaps cut closest to the nub. Most colonists were simply not interested in what inventors were doing, as Hobday, Colles, and others found out. Those colonists who were had a limited frame of reference.

They had, however, charted themselves a new course. By expanding their political aspirations, they put greater pressure on the economy. The greater the pressure they put on the economy to sustain their political ambitions, the more pronounced became the need for technological change. Americans after the war would continue searching for ways to achieve technological independence, and in the process they would learn just how technologically limited they really were, given their ambitions. What is more, after the war they would come to see inventors and inventions as essential to technological change, just as they saw technological change in manufacturing as the answer to their most basic economic woes.

Equally important, the public would eventually realize that inventors did not act alone. Tully and Hague may have been able to make machines for the United Company of Philadelphia, yet the company collapsed, their inventiveness notwithstanding. Machines alone could not compensate for the lack of capital; inventiveness could not offset constricted markets. The colonies had produced no Watt of their own, in part because they were not at the technological point where Watt machines could be applied. Watt himself could not have single-handedly wrought sudden, successful change. Only when other factors were in hand would the inventor become a catalyst for techonological change. Until that time inventors remained, to the public, at least, tinkerers, masters of the curious; until that time inventors remained, much to their sorrow, on the fringe of American life.

As an inventor in Colonial America would have been the first to remark, many forms of ingenuity, however worthy, were virtually ignored. Most successful technologies had been adaptive, recreating in the New World, with some important modifications, the technologies of the Old World. The inventor had not yet become a folk hero because he had not yet proved himself. Colonial men of science, such as those belonging to the American Philosophical Society, had a better feel for invention than did the general public. But even they did not see invention the way they would after the war, after technological transfer from Britain brought the beginnings of a new industrialization, an industrialization destined to fire the national imagination.

The colonists were not futurologists, and the technological shortcomings exposed during the war had not always been evident before it. Furthermore, the war imposed technological demands non-existent previously, stretching what had been adequate during peace to the breaking point and beyond. If the war taught political lessons, if it served as a testing ground for ideas about representation, right, and responsibility, it also taught costly technological lessons. Manufacturing deficiencies during the war, even more than those before it, would produce a shock, charging with renewed energy the desire for technological progress.

During the War of Independence the patriots fought as might be expected of a provincial, pre-industrial society. They did not miraculously marshal their forces; they did not draw on some innate though hidden technological expertise to radically alter their economy and restructure their politics. During the war their technological performance was uneven, as it had been in the pre-war years. They did well in some areas, less well in others; the more complex the undertaking, the greater the chance for failure. American technology was in its adolescence. Nonetheless, it was also undergoing a change, a metamorphosis paralleling the revolution in American politics.

NOTES

1. Bridenbaugh, *Colonial Craftsman*, p. 6. Consult Brooke Hindle's *Technology in Early America* (Chapel Hill: University of North Carolina Press, 1966) for a guide to studies of craftsmen and their work.

2. See Morris, *Government and Labor,* pp. 24-25, 27, 28; Jeremy, "British Textile Technology Transmission to the U.S.," p. 43; and A. E. Musson and Eric Robinson, *Science and Technology in the Industrial Revolution* (Toronto: University of Toronto Press, 1969), pp. 216-230, for the inconsistent secrecy of Matthew Boulton.

3. *New York Journal,* 24 December 1767.

4. *Va. Gazette* (Hunter), 6 June 1751.

5. *Pa. Gazette,* 15 November 1770; also 16 March and 1 June 1769 for other English flaxdressers.

6. *Boston Gazette,* 9 September 1771; *New York Journal,* 23 October 1766.

7. *Boston Gazette,* 27 May 1765, supplement.

8. *New York Journal,* 8 October 1767.

9. *Pa. Gazette,* 1 May 1766 and 26 October 1769.

10. Manufacturer in ibid., 30 January 1772; machinery in the *New York Journal,* 12 January 1775; also the *Va. Gazette* (Pinckney) 19 January 1775 for English manufacturers in Newport, Rhode Island; Virginia in ibid., 17 October 1774.

11. Mobs noted in the *Pa. Gazette,* 22 October 1767; and James L. Bishop, *A History of American Manufactures, 1608-1860,* 2 vols. (Philadelphia: Edward Young and Co., 1864), II, 93-94; opposition to Hargreaves is mentioned in Paul Mantoux, *The Industrial Revolution in the Eighteenth Century* (1928; reprint ed., New York: Harper and Row, 1961), p. 218, and in most histories of the Industrial Revolution. But as Phyllis Deane, *The First Industrial Revolution* (Cambridge: Cambridge University Press, 1965), pp. 147-151, makes clear, such acts did not represent a class revolution by factory workers. Indeed, the "factory system" was not yet in place, even in textiles, and would not be for another two decades.

12. Smyth, ed., *Writings of Franklin,* VI, 139; also Franklin to Humphrey Marshall, 22 April 1771, in the Simon Gratz Autograph Collection, Case 1, Box 19, Historical Society of Pennsylvania Manuscripts (hereafter HSP MSS.). Drew R. McCoy, *The Elusive Republic* (Chapel Hill: University of North Carolina Press, 1980), pp. 48-75, notes Franklin's opposition to large-scale manufactures and the difficulty of reconciling manufactures and commerce with "classical republicanism." I am not convinced, however, that Franklin was all that preoccupied with the issue, nor do I think that the "conflict" between commerce and republicanism was all that clear to Franklin's contemporaries. In my view, McCoy pushed his arguments too far. Franklin had invested in a Philadelphia linen factory that opened in 1764, only to close three years later (Labaree, ed., *Papers of Franklin,* XI, 314-316, 316n). He was genuinely disappointed that the experiment failed.

13. *Journal of Burgesses,* X, 362; *New York Journal,* 15 January 1767; and the *Va. Gazette* (Purdie), 4 July 1766. See Michael Kraus, *The Atlantic Civilization* (Ithaca: Cornell University Press, 1949), pp. 172-175, in particular.

14. *Va. Gazette* (Purdie), 15 September 1775, supplement; and ibid. (Purdie and Dixon), 7 November 1773; and Schlesinger, *Colonial Merchants,* pp. 517-518.

15. Notes on activities in the *Pa. Journal,* 25 April 1765; the *Pa. Gazette,* 2 August 1764 and 7 February 1765; the *Va. Gazette* (Purdie and Dixon), 6 July 1769 and 13 June 1771; and elsewhere in these and other papers.

16. See for example the notes in the *Pa. Gazette,* 6 September 1764; and the *Va. Gazette* (Purdie and Dixon), 29 August 1771.

17. Great Britain, Series B, American Correspondence and Transactions, Royal Society of Arts, Guard Book, A91 (reel 1); Guard Book, IX, 119 (reel 2), LC microfilm.

18. Ibid., Guard Book, IX, 127 (reel 2), LC microfilm.

19. Ralph Bates, *Scientific Societies in America* (1945; reprint ed., New York: Columbia University Press, 1958), pp. 4-9; and Brooke Hindle, *The Pursuit of Science in Revolutionary America, 1735-1789* (Chapel Hill: University of North Carolina Press, 1956), pp. 127-145, discuss the competition between two rival societies leading to the formation of the American Philosophical Society, and both provide an overview of the society's membership, goals, and effectiveness. Also see Brooke Hindle, "The Rise of the American Philosophical Society, 1766-1787" (Ph.D. dissertation, University of Pennsylvania, 1949), passim.

20. *Pa. Gazette*, 17 March 1768.

21. "Early Proceedings of the American Philosophical Society for the Promotion of Useful Knowledge, 1744-1838," *American Philosophical Society Proceedings* 22 (July 1885), part 3, no. 119, p. 30; William Smith, *An Oration* (Philadelphia: John Dunlap, 1773), p. 8.

22. Benjamin Franklin, *Proposals Relating to the Education of Youth in Pennsylvania* (Philadelphia: Franklin and Hall, 1749), pp. 28-29. Franklin also stated that the study of history in general should develop an appreciation for how the "Arts [are] invented, and Life made more comfortable" (p. 22). After all, he asked rhetorically, "How many Mills are built and Machines constructed, at great and fruitless Expense, which a little Knowledge in the Principles of Mechanics would have prevented?" (p. 28n). For the APS circular letter of May 14, 1743, see Labaree, ed., *Papers of Franklin*, II, 381-382.

23. Francis Hopkinson, *Science: A Poem* (Philadelphia: Andrew Steuart, 1762), p. 1.

24. *Pa. Gazette*, 28 March 1771.

25. Ibid., 19 December 1771. Inventions noted in the *APS Trans.* 1 (1771) include William Henry's "Sentinel Register for regulating the flame of a furnace" (pp. 286-289); Richard Wells's ship pump (pp. 289-292); Owen Biddle's file cutting machine (pp. 300-302); and Thomas Gilpin's "Horizontal Wind-Mill" for raising water out of mines and wells (pp. 339-340).

26. For the involvement of Philadelphians, see Whitfield Bell, "The Scientific Environment of Philadelphia, 1775-1790," *APS Proceedings* 92 (1948):6-14; idem, "Science and Humanity in Philadelphia, 1775-1790" (Ph.D. dissertation, University of Pennsylvania, 1947), passim; Carl and Jessica Bridenbaugh, *Rebels and Gentlemen* (New York: Reynal & Hitchcock, 1942), p. 305; and a different interpretation of public interest and participation in science at this time in Dirk Struik, *Yankee Science in the Making* (Boston: Little, Brown and Co., 1948), pp. vii-viii.

27. Philip S. Foner, ed., *The Complete Writings of Thomas Paine*, 2 vols. (New York: Citadel Press, 1945), II, 1024; from Paine's "Useful and Interesting Hints," published in the *Pennsylvania Magazine* 1 (February 1775):57.

28. *Pa. Gazette*, 8 March, 15 March and 22 March 1770; 31 January, 19 September and 3 October 1771; 9 June 1773; and 30 March 1774. Also see the "Early Proceedings of the APS," p. 64.

29. "Early Proceedings of the APS," pp. 50, 60; and the *Pa. Gazette*, 31 March 1773. APS member Timothy Matlack owned a brewery, while member John Rhea ran a pot and pearl ash operation.

30. See ibid., pp. 32, 34, 35, 53; *Chesapeak* (Philadelphia, 1768); John F. Watson, *Annals of Philadelphia and Pennsylvania*, 3 vols. (1836; revised by Willis Hazard, Philadelphia: Edwin Stuart, 1900), II, 466-468; Miller, *Retrospect*, I, 374-375; and

two anonymous broadsides, "To the Public" (15 January 1772) and "A Friend to Trade" (13 December 1771). See the *Pa. Gazette*, 15 August 1771; and also ibid., 20 February 1772, for interest in turnpikes. Franklin recommended that a British engineer be recruited and paid a "handsome salary" for internal improvements: see his letter to Samuel Rhoads of 22 August 1772 in Labaree, ed., *Papers of Franklin*, XIX, 278.

31. *Va. Gazette* (Purdie and Dixon), 13 May and 22 July 1773; also Silvio A. Bedini, *Thinkers and Tinkers* (New York: Charles Scribner's Sons, 1975), pp. 181-182.

32. *Va. Gazette* (Purdie and Dixon), 5 August 1773.

33. Morris, *Government and Labor*, p. 23; Tryon, *Household Manufactures*, pp. 26-27; also see the contemporary notices in the *Pa. Gazette*, 10 January 1765; and the *Va. Gazette* (Purdie and Dixon), 6 and 13 October 1774. The Virginia Assembly in 1760 had formed a committee—Peyton Randolph and George Wythe were among the members—to correspond with the Royal Society of Arts. See Great Britain, Series B, American Correspondence and Transactions, Royal Society of Arts, Guard Book, II, 46, 50 (reel 1), LC microfilm.

34. Roger Burlingame, *March of the Iron Men* (New York: Charles Scribner's Sons, 1938), p. 63, implied that Americans were more aware of the importance of invention than they really were, or possibly could have been; Struik, *Yankee Science*, pp. 8-9, though more restrained than Burlingame, said essentially the same thing. See Meier, "Technological Concept," pp. 14-18, for a more balanced assessment.

35. Canals in the *Va. Gazette* (Purdie and Dixon), 24 August 1769 and 28 April 1774; ship's pump in the *Pa. Gazette*, 26 January 1764; spinning wheel in the *Va. Gazette* (Purdie and Dixon), 16 March 1769; ibid., 22 August 1771 for desalinization; hydrometer in the *Pa. Gazette*, 18 April 1773; windlass in the *Va. Gazette* (Purdie and Dixon), 18 November 1773; and ibid., 15 October 1767 for dyes; *Pa. Gazette*, 7 February 1760, for copper sheathing to coat ship bottoms.

36. Musical chair in the *Pa. Journal*, 2 May 1765; itch cure in the *Boston Gazette*, 2 January 1764; leather hardener in the *Va. Gazette* (Purdie and Dixon), 21 April 1768; steel softener in ibid.

37. *Pa. Gazette*, 4 March 1762; *Va. Gazette* (Purdie and Dixon), 25 February 1768.

38. *Va. Gazette* (Purdie and Dixon), 24 August 1769, for rocket boat; ibid., 29 June, 28 September and 23 November 1769, and 26 September 1771, for horseless carriages; ibid., 27 August 1772 for the submarine. See ibid., (Hunter), 18 April 1751 and 20 March 1752 for reports on a perpetual motion machine and a glider made out of cork and feathers.

39. Carroll W. Pursell, Jr., *Early Stationary Steam Engines in America* (Washington, D.C.: Smithsonian Institution Press, 1969), pp. 4, 10. Brooke Hindle has noted that "the transfer of technological systems represents an achievement of a different order of magnitude from the transfer of a single invention," as cited in "The Transfer of Power and Metallurgical Technologies to the United States, 1800-1880," *Colloques Internationaux*, no. 538, p. 408.

40. *Boston Gazette*, 13 February 1764; also see the complaint of Thomas Abel that the colonists were not really interested in promoting the "useful arts" in the *Pa. Gazette*, 29 October 1761. Abel was upset because subscribers were not paying for or picking up their copies of his book on trigonometry.

41. *Pa. Gazette*, 22 August 1771.

42. For a note on the beneficial and detrimental character of patents, see Ashton, *Industrial Revolution*, pp. 10-11.

43. Hindle, *Technology in Early America*, p. 91. For colonial patents see F. W. Dahn, "Colonial Patents in the United States of America," *Journal of the Patent Office Society* 3 (1921):342-349; P. J. Federico, "Colonial Monopolies and Patents," *Journal of the Patent Office Society* 11 (1929):358-365; Frank D. Prager, "Historic Background and Foundation of American Patent Law," *The American Journal of Legal History* 5 (1961): 309-325; and Bruce W. Bugbee, *Genesis of American Patent and Copyright Law* (Washington, D.C.: Public Affairs Press, 1967), pp. 12-83.

44. *The Book of the General Lawes and Libertyes Concerning the Inhabitants of the Massachusets* (Cambridge: Samuel Green, 1660), p. 61; *The Book of the General Lawes For the People within the Jurisdiction of Connecticut* (Cambridge: Samuel Green, 1673), p. 52; and Nicholas Trott, ed., *The Laws of the Province of South Carolina*, 2 vols. (Charles Town: Lewis Timothy, 1738), I, 21. The South Carolina act was declared "obsolete" by 1736.

45. Nathaniel B. Shurtleff, ed., *Records of the Governor and Company of the Massachusetts Bay in New England, 1628-1686*, 6 vols. (Boston: William White, 1853-1854), II, 149. Federico, "Colonial Monopolies and Patents," pp. 360-361; and Dahn, "Colonial Patents," pp. 344-345, point out that the first patent went to Samuel Winslow in 1641 for a saltmaking process.

46. Charles J. Hoadly, ed., *The Public Records of Connecticut*, 15 vols. (Hartford: Case, Lockwood and Brainard Co., 1890), VII, 174; X, 231; XIV, 244.

47. Samuel Hazard, ed., *Pennsylvania Colonial Records* (Harrisburg: Historical Society of Pennsylvania, 1852), III, 18-19; and *Minutes of the Provincial Council of Pennsylvania*, 10 vols. (Philadelphia: Joseph Severns and Co., 1852), I, 34.

48. British Museum, King's MSS no. 206, pp. 60-61, LC.

49. Bridenbaugh, *Colonial Craftsman*, p. 56.

50. *Pennsylvania Magazine* 1 (April 1775):158; Bishop, *American Manufactures*, I, 394-395; and Jeremy, "British Textile Transmission to the U.S.," pp. 28, 32, 40.

51. "Early Proceedings of the APS," pp. 12-13, 39, 66, 68, 77, 80. John Jones's letter of 26 October 1771, describing his scythe machine, and a letter of 27 May 1773, describing an impractical suspension bridge for the Schuylkill River, are in the APS archives.

52. Description and illustration in the *Pennsylvania Magazine* 1 (May 1775):206-208.

53. "Early Proceedings of the APS," pp. 78, 80, 82, 97.

54. Samuel Hazard et al., eds., *Pennsylvania Archives* (Harrisburg: Historical Society of Pennsylvania, 1852-), 8th series, VIII, 7189, 7191.

55. McIlwaine and Kennedy, eds., *Journals of Burgesses*, X, 283, 294-295, 301.

56. *Va. Gazette* (Purdie and Dixon), 19 November 1772; 28 January, 27 May and 2 December 1773.

57. McIlwaine and Kennedy, eds., *Journals of Burgesses*, XIII, 111-112, 116-117; *Va. Gazette* (Purdie and Dixon), 16 June 1774.

58. *Va. Gazette* (Rind), 26 April 1770, supplement; ibid. (Purdie and Dixon), 20 February 1772.

59. *Pa. Gazette*, 21 July 1768; imitation in ibid., 22 August 1771.

60. Ibid., 19 November 1767; see ibid., 10 December and 17 December 1767,

for an attack on Wells, charging him with copying the spring from an English source, and Wells's stinging reply. Wells became one of the twelve managers of the United Company of Philadelphia factory in 1775.

61. Ibid., 13 October 1768; *New York Journal*, 17 August 1769, supplement.

62. "Early Proceedings of the APS," p. 85, 90; and Wells, "Account of a snow plough," APS archives, 17 December 1773. Wells served as APS secretary from 1774 to 1776 and after the war was involved in the Fitch-Rumsey steamboat controversy.

63. Walter Ristow, ed., *A Survey of the Roads of the United States of America* (Cambridge: Harvard University Press, 1961), p. 3; and John Austin Stevens, "Christopher Colles: The First Projector of Inland Navigation in America," *Magazine of American History* 2(1878):340-348.

64. *Pennsylvania Chronicle*, 26 August 1771; also see the letter of introduction from John Murgatroyd of Philadelphia in the HSP Misc. MSS. Collection.

65. Hazard, ed., *Pa. Archives*, 8th series, VIII:6929.

66. *Pa. Gazette*, 29 September and 10 November 1773.

67. "Early Proceedings of the APS," pp. 82, 83.

68. George W. Edwards, *New York as an Eighteenth Century Municipality* (New York: Columbia University Press, 1917), pp. 139-141, gives too little space to Colles, but is informative in other areas. Also see the description of the system in Edward Bangs, ed., *Journal of Lieutenant Isaac Bangs* (Cambridge: John Wilson and Son, 1890), pp. 25-26. Notes are also in the *Va. Gazette* (Purdie), 3 March 1775, supplement; Pursell, *Stationary Steam Engines*, pp. 7-9; and Ristow, ed., *Survey of the Roads*, pp. 13-14. I am indebted to Brooke Hindle for a copy of his paper on "Water Supply Systems in Early America," which compares and contrasts the Colles experiment in Manhattan with the Philadelphia waterworks of Benjamin Henry Latrobe and the earlier Bethlehem water supply system.

69. *Pa. Gazette*, 1 August 1771.

70. Labaree, ed., *Papers of Franklin*, V, 526-527.

3 WAR ECONOMY: THE MUNITIONS INDUSTRY

According to I. Bernard Cohen, with the outbreak of war in April 1775, American revolutionaries faced three basic problems related to science and technology: creating a war industry, establishing a medical department, and recruiting European military engineers.[1] The first was without question the most pressing of the three. Congress, the states, and local committees of safety had to build their war industry from the ground up. Most historians agree that the war fostered industrial growth to a much greater extent than had the home manufactures movement of the preceding decade. Hugo Meier wrote that the Revolution "stimulated industry as had no other factor" before it, and a number of economic historians concur with his view.[2] Robert East, for instance, after examining the rise of new business groups and the expansion of markets, concluded that "there is considerable evidence of a gainful spirit at work during the Revolution." Nevertheless, these same historians also agree that industrial growth was not uniform, and investment in heavy manufactures was lacking.[3]

The munitions industry provides a good case study of the wartime manufacturing experience as a whole. Inadequacies in the munitions industry, on both production and managerial levels, were symptomatic of deficiencies in the entire industrial sector. In most cases these inadequacies had not been anticipated. Because of the quantity and accessibility of iron ore deposits, the presence of skilled workers, and the familiarity of most white male colonists with firearms, leading patriots at first supposed they would have no difficulty producing adequate munitions stocks. A Pennsylvanian claimed, "There are . . . gunsmiths enough in this province to make 100,000 stand of arms in one year . . .

Part of the material which follows first appeared as an article in *Military Affairs*, February 1979, pages 26-30. Reprinted with permission.

if they should be wanted."[4] John Adams boasted in February 1775 that the colonists had abundant arms and ammunition, and "if we had not we could make a sufficient quantity of both."[5]

The demands of full-scale war soon proved them wrong. Powder mills had been built during the French and Indian War, but no more than one or two were operating in April 1775. The rest had been shut down; many rotted in disrepair. The excitement of pre-war home manufactures brought no significant change, and this moribund industry could not be reinvigorated overnight. Besides, rebuilding mills would not alone answer the need, for Americans had to transform what had traditionally been a localized system of production into something quite complex. Munitions production involved much more than the simple making of gunpowder or assembling of muskets. That gunpowder and those muskets also had to be delivered to troops in the field. Production and delivery depended on a complicated network as political as it was economic. The British produced munitions in much the same manner as the patriots—by contract through small shops—but they did so with fewer problems. Despite all endeavors and claims to the contrary, the patriots never did become self-sufficient. Wartime stimulation and expansion of the munitions industry did occur, but not to a degree allowing Americans to fend for themselves, without foreign aid.

Admittedly, a handful of prescient colonists had foreseen the coming of war and the need for manufacturing munitions at home as well as purchasing them abroad. In 1770 a British merchant warned Lord Hillsborough, the colonial secretary, that Bostonians were arming.[6] While this warning may have been premature, some patriots definitely prepared for hostilities months before they actually erupted. In December 1774 a few hundred colonists absconded with the munitions kept at Fort William and Mary in Portsmouth, New Hampshire, while patriots in Massachusetts secretly stockpiled munitions in towns ringing Boston.[7]

An October 1774 Parliamentary edict forbidding exportation of munitions of any sort to the colonies raised a cry of protest, causing still more patriots to prepare for the worst. The *Royal American Magazine* cautioned its readers that Parliament proved it wanted "to make us an easier prey to the vulture jaws of tyranny."[8] The editors appended a recipe for making saltpeter (or potassium nitrate), a key ingredient of gunpowder. At approximately the same time, Rhode Islanders began making firearms; they had been gathering gunpowder and lead for some time.[9] Many committees of safety around the colonies left loopholes in their non-importation resolutions to allow for the continued gathering of the raw materials of war.

By and large, however, only a tiny amount of munitions had been manufactured in the colonies or slipped in from Europe before the war. American revolutionaries had little more than the small quantities sequestered in gunshops, trading houses and private stocks around the country. Indeed, General William Moultrie reminisced that he and his fellow patriots had dared to oppose Great Britain "without money; without arms; without ammunition; no generals; no armies; no admirals; and no fleets." At the time of Lexington and Concord, Moultrie's home state of South Carolina stood defenseless. South Carolinians seized an English brig carrying some gunpowder, yet "the want of powder was a very serious consideration with us," recalled Moultrie, for "we knew there was none to be had upon the continent of America."[10]

Provincial conventions and committees of safety sought to relieve the shortages. Subsequent price controls and production quotas smacked of mercantilism to some malcontents, but such legislation was the only effective way of building an industry almost non-existent before the war. Each province implemented a munitions production program, with some doing much more than others. Craftsmen were diverted from peaceful pursuits—bricklaying, clockmaking, carpentry—to run powder mills and assemble firearms. Most worked under a system of contracts with local committees of safety or with commissioners appointed by the Continental Congress.

New England, for example, worked through existing authorities, primarily town selectmen and local committee of safety members.[11] Beginning in May 1775, Connecticut enacted legislation designed to stimulate the munitions industry there. The General Assembly appointed seven men to oversee firearms production and promised a bounty of five shillings to gunsmiths for each stand of arms made according to their specifications. The assembly called for three thousand muskets stamped "with the maker's name" and raised the total some months later. Bounties were also granted for saltpeter and sulphur, which were combined with charcoal to produce gunpowder. Each town in the province sending a representative to the assembly and lacking any privately owned saltpeter mills was directed to erect one under the supervision of town selectmen. All mills, public or private, were to deliver their saltpeter to powder mills. The powder mills, after marking each cask clearly with the weight and maker's initials, deposited the finished product with the provincial committee of safety. Selectmen also had charge of stockpiling lead, with authority to confiscate whatever they needed.[12] Late in 1776 the assembly appointed inspectors to check on the town selectmen and make the rounds to each saltpeter and powder mill.[13] Privately owned mills in New Haven, Windham, and Norwich operated successfully under this arrangement. Lead mines at Middletown, New Canaan, Farmington, and Northampton enjoyed marginal success. The province also financed construction of a furnace for cannon and shot at Salisbury and powder mills where necessary.[14]

Massachusetts did much the same. It sought to further production

WHEREAS it is of the utmost Importance to the Welfare and Happiness of these Colonies, that the manufacturing of Fire Arms and Provisions of Military Stores, be effectually promoted and encouraged, agreeable to the Recommendations of the honorable Continental Congress: Therefore,

Resolved, That for every effective and substantial Fire Arm, which shall be manufactured in this Colony, with a Barrel of three Feet and nine Inches in Length, that will carry an Ounce Ball, a good Bayonet, with a Blade not less than eighteen Inches in Length, a Steel Ramrod, with a Spring to retain the same, two Loops for Gun-Strings, and the Maker's Name stamped or engraved on the Lock, and which shall be delivered at *Watertown,* to *Richard Devens,* Esq; Commissary, on or before the first Day of *June* next, and resemble in Construction, and as nearly as may be equal in Goodness the King's new Arms, there shall be allowed and paid out of the public Treasury to the Owner thereof, the Sum of *three Pounds;* and for the Accommodation and Convenience of such Manufacturers, it is also *Resolved,* That Col. *John Baker,* of *Ipswich,* in the County of *Essex;* Capt. *Ichabod Goodwin,* of *Berwick,* in the County of *York;* Capt. *Oliver Witt,* of *Paxton,* in the County of *Worcester;* Capt. *Elijah Hunt,* of *Northampton;* Mr. *Aaron Ashley,* of *Springfield,* in the County of *Hampshire;* Capt. *David Bush,* of *Pittsfield,* in the County of *Berkshire;* Dr. *David Cobb,* of *Taunton,* in the County of *Bristol;* Mr. *Michael Blackwell,* in the County of *Barnstable;* and Capt. *James Hatch,* of *Pembroke,* in the County of *Plymouth;* be, and they hereby are respectively impowered to receive for the Colony, at the Price and during the Time aforesaid, all Fire Arms which shall be offered them for Sale, and manufactured as aforesaid; and they are directed to deliver the same once a Month to the Commissary, and taking his Receit therefor, to apply to the Council for a sufficient Sum wherewith to pay for such Fire Arms, the Charges of Transportation, &c.

Provided always, That the Owner of each Fire Arm which shall be received for the Colony, shall prove the same at his own Risque, by four Inches and a half of Powder, a Ball and Wads on each in Presence of one of the Committee aforesaid, or in Failure thereof, before the Commissary. And it is hereby recommended to the Committee of Correspondence, and Selectmen of each Town in this Colony, to apply to the Manufacturers of Fire Arms in their respective Towns, and afford them all necessary Encouragement, and to post this Resolve, that the Inhabitants of each Town may promote this salutary Measure.

Sent up for Concurrence.

J. WARREN, Speaker.

In Council, *November* 4, 1775.

Read and Concurred. REREZ MORTON, Dep. Sec'y.

Consented to,

JAMES OTIS,	*M. FARLEY,*
W. SPOONER,	*B. LINCOLN,*
CALEB CUSHING,	*J. PALMER,*
JOSEPH GERRISH,	*JABEZ FISHER,*
JOHN WHETCOMB,	*S. HOLTEN,*
JED. FOSTER,	*MOSES GILL,*
JAMES PRESCOT,	*B. WHITE.*
ELDAD TAYLOR,	

A true Copy, Attest, PEREZ MORTON, Dep. Sec'y.

6. Massachusetts munitions production notice, detailing the way in which arms were to be gathered. Courtesy of the Massachusetts Historical Society.

with bounties yet maintain rudimentary quality control by granting premiums only to those gunshops and powder mills following official specifications. The provincial assembly assessed quotas for each town and county, and selectmen once again were expected to keep things running smoothly.[15]

Initially New York was not so well organized. Its provincial government expressed a willingness to indiscriminately "hire all the ARMS, with or without Bayonets, that are fit for Present Service." Yet the New York Congress wisely sent gunsmith Henry Watkeys on an expense-paid trip to New Windsor, Connecticut, to talk with gunsmiths there about making muskets.[16] And in March 1776, New York offered interest-free, two-year loans to anyone building a powder mill or gunlock factory. As added incentives it promised bonuses up to £100 for the first three powder mills and gunlock factories, as well as bonuses to gunsmiths making a certain number of muskets, this time according to government specifications.[17] Mills along the Hudson River churned out gunpowder, with Henry Wisner of Ulster County alone processing about nine thousand pounds over a three-month period in 1776.[18]

Pennsylvania moved energetically, having by November 1775 appointed Robert Towers as chief of commissary. The committee of safety appointed commissioners to contract with local gunsmiths for one thousand firearms.[19] In February 1776 the committee erected a gun manufactory in Philadelphia run by Peter De Haven, and offered private entrepreneurs loans of up to £150 and bonuses of up to £100 to build saltpeter and powder mills. Berks and Lancaster counties led the province in firearms production, but gunsmiths in Northampton County produced nearly as many muskets and rifles. Furnaces at Reading, Warwick, and Carlisle cast cannon.[20]

Virginia, if anything, moved even more quickly than Pennsylvania. In July 1775 the Virginia committee of safety passed an ordinance for supplying arms and ammunition to the provincial government. A manufactory for firearms was set up at Fredericksburg, run by state-appointed commissioners and operated at public expense.[21] The government invited qualified craftsmen and apprentices to apply for employment at the factory. It also resolved to award premiums for saltpeter and to compensate the first two individuals setting up plating and slitting mills for any losses they might sustain.[22] In December, Virginia also contracted with private gunsmiths and made provisions for publicly owned powder mills. The state later rented a lead mine in Fincastle County and granted James Hunter a parcel of land to supply his forge with iron ore.[23] Like Virginia, South Carolina subsidized its home industry. It granted a premium of £1000 for the first bloomery to make one ton of iron and promised smaller bounties for saltpeter and sulphur mills, and a gunlock factory.[24]

The states were not left to their own devices. The Continental Congress was equally concerned with promoting home munitions production. Congress started out gingerly, recommending that the states (technically still colonies) decide on standard requirements for their firearms. It also encouraged them to erect powder mills.[25] Congress eventually graduated beyond this stage when its members realized that they had to do more than advise the states and show their backing of American manufactures by parading about Philadelphia in homespun. Yet they did so only slowly, waiting until February 1776 to assemble a committee of five to consider ways of promoting munitions production.[26]

Congress very early reached its limits. In March 1776 it chose not to try to take charge of manufactures, including the munitions industry, nationwide, most likely to the disappointment of John Adams. Adams had hoped that Congress would go further. He headed a committee recommending that each province form a society "for the Encouragement of Arts, Manufactures, Agriculture and Commerce." Adams's colleagues went along with the recommendation, but only in part. An all-important clause providing for a standing congressional committee to correspond with and assist those societies was erased, leaving Adams's resolution incomplete and ineffectual.[27] Congress would come no closer to assuming leadership. It remained reluctant to go beyond an advisory role, even after the Declaration of Independence and its emergence as the national government.

It did take at least symbolic command of the munitions industry by creating the Board of War (and Ordnance) in June 1776, almost six months after an agency of this type had been suggested and adumbrated.[28] During its five-year tenure, the Board of War continually broadened the scope of its duties. It appointed inspectors to examine each barrel of gunpowder collected for Continental use, with inspected barrels stamped "USA." To "prevent abuses that have been so much practised by Gunsmiths and others employed in public Arms," the board designated Thomas Butler as the "Public Armourer" in 1777 to superintend "the repairing and proving of all public Arms."[29] Shortly thereafter Congress set aside funds for two magazines, one at Springfield, Massachusetts, the other at Carlisle, Pennsylvania. The Board of War was put in charge of both magazines.

Henry Knox had recommended this arrangement back in September 1776 in a report he called "some hints for the improvement of the Artillery of the United States." He suggested that Congress erect two "Laboratories" and supporting foundries for manufacturing, proving, and storing munitions.[30] Washington consequently authorized Benjamin Flowers to organize a regiment of "artificers" for the installations proposed by Knox and instituted by Congress. The regiment was composed of companies of craftsmen—wheelwrights, blacksmiths, carpenters, harness-makers, coopers, and nailers—who either manned the "Laboratories" or marched with the army.[31]

In addition to enlisting Flower's regiment, Congress joined the states in granting special deferments to skilled craftsmen. In September 1777 it directed Washington to detach from the army a sufficient number of workers to repair two to three thousand damaged muskets gathering dust in Philadelphia. Congress regretted that "there is no other mode of supplying the Demands" of unarmed militiamen in Pennsylvania, Delaware, and Maryland.[32] Congress had earlier exempted eleven Pennsylvanians employed at a cannon foundry, "as it is represented that the Works must stand idle if these Workmen" joined the fighting. Lancaster County, Pennsylvania, exempted numerous gunsmiths, weavers, and others from service as well.[33]

Government officials received aid from newspapers, magazines, and almanacs publicizing the need for munitions. Broadsides, pamphlets, newspaper columns, and magazine articles authored by Franklin, Thomas Paine, Benjamin Rush, and John Winthrop of Harvard gave step-by-step instructions on the making of saltpeter and gunpowder. The authors appealed to their readers' patriotism and civic voluntarism. Stressing the desirability of self-sufficiency because the war threatened normal commercial relations with Continental Europe, the Pennsylvania and New York committees of safety sponsored the printing of pamphlets on making gunpowder and included pieces by Franklin and Rush. Rush had earlier described the process for readers of the *Pennsylvania Gazette*, and that paper also served notice of the time and place of technical demonstrations. (Royal authorities in Pennsylvania had undoubtedly looked askance at Rush's pseudonymous articles on the making of saltpeter, which began appearing as early as November 1774.)[34] *Bickerstaff's New England Almanac* for 1776 squeezed gunpowder-making directions along the top margin of its meteorological tables. The editors admonished their readers that gunpowder could be processed "by almost every Farmer in his own Habitation."[35]

Everywhere Americans turned they were exhorted to contribute to the cause. They were urged to make the armaments industry an impressive display of American might, but they were not able to do as they were asked. Despite enthusiastic backing from various quarters, the munitions industry stumbled over obstacle after obstacle. The basic problem lay within the system itself. Though most states set up minimal requirements, product quality differed markedly, depending on the comparative efficiency or laxity of enforcement and the abundance of raw materials. Strict adherence under the decentralized programs run by state committees of safety proved virtually impossible. In all too many cases committeemen had to take what they could get because they could ill afford to be choosy.

Many states found it impossible to follow through on pledges to erect publicly owned manufactories. In spite of its best-laid plans Virginia did not have a single publicly financed powder mill by April 1776.[36] Maryland's attempt to establish a state-run manufactory failed, and it had to rely exclusively on munitions purchased from private parties. Four months after the war had started, South Carolina lacked a single powder mill, public or private. Gunsmiths were widely distributed across Colonial America, so initial problems with firearms were not as acute. Ample supplies of wood for stocks and iron for barrels enabled most gunsmiths to avoid the resource deficiencies that harried many powder makers. Problems there were, however. Lancaster County, Pennsylvania, gunsmiths complained that other counties had better contracts, so they slowed production until their grievances were met and fell behind on their quota as a result. At one point Lancaster gunsmiths became so uncooperative that the county committee of safety had to order some of them not to leave their homes or engage in any other work until they filled their contracts.[37]

Massachusetts patriots had similar problems. Armorers had been appointed by the provincial committee of safety within a week of the skirmish at Concord, but even after the passage of several months they were not keeping the troops adequately supplied. Many of the weapons they turned in for service were unsuitable, despite the temporary willingness of the provincial congress to take virtually anything it could get. As in Pennsylvania, the mixture of voluntarism and government contract attempted by Massachusetts did not satisfy demand. A like scenario could be painted for the other colonies.[38]

Variations in caliber, length, "furniture," and overall quality in committee of safety muskets were unavoidable, a result of the hazy conception of standardization in a pre-machine tool age. While some muskets were patterned after British models like the "Brown Bess," others were patterned after French weapons, while still others were not patterned after any particular design. Massachusetts called for muskets that measured three feet ten inches long with a sixteen-inch bayonet. Connecticut called for muskets one inch shorter and bayonets at least two inches longer. The Continental Congress called for muskets yet another inch shorter. New York simply requested muskets "at least" three and a half feet long with "good bayonets." Committees of safety ambiguously requested "suitable" gunlocks and "substantial" triggerguards and paved the way for deviations defeating efforts at standardization. Emphasis on filling as many orders as possible, as soon as possible, may in turn have adversely affected serviceability. Inspectors assigned to oversee production frequently relaxed their attentiveness, further diminishing the application of rigorous standards.[39] As a result, some gunsmiths were guilty of "making poor, deficient arms, totally unfit for service." Robert Treat Paine grumbled that more care had to be taken with the manufacture of gunpowder because some "miserable trash" had been turned out.[40] Even under the best conditions, production moved at a snail's

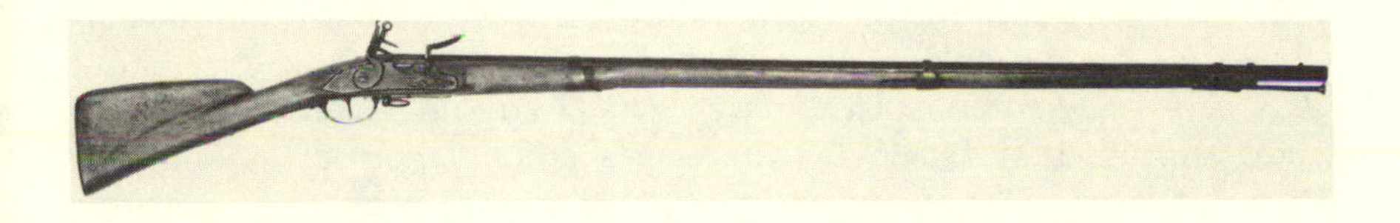

7. A committee of safety musket, copied from a 1754 French model. Smithsonian Institution Photo No. 69038.

pace. Small shops employing a master craftsman and one or two apprentices could complete about twenty guns per month, if that.

On the national level, Congress had no way of anticipating the amount of munitions the army would require until it had weathered a few campaigns. And accurate computation was complicated by the fact that there was not really a single, unified American army. The American military establishment was divided into militia companies, state regiments, and Continental units, and never became truly national. States had quotas to fill for each group. Congress, in addition to keeping track of those activities, struggled to collect its own munitions reserve, administered and dispensed by the Board of War and the commissary of military stores. Because the Board of War served as an adjunct of Congress, its coercive power was negligible. Much to its chagrin, it found that repairing weapons damaged by long service or misuse was as difficult as obtaining new guns. Sometimes it had to compete with states, counties, and even towns for precious supplies. Undoubtedly one or two states hoarded stocks, unwilling to strip their militia of arms and ammunition even though they might be far removed from the scene of war. Congress could do nothing to punish or coerce them. Paralyzed to an extent by ill-defined powers, neither the Board of War nor Congress could orchestrate a closely managed program.

Short-lived and generally impotent manufacturing societies, begun in a few states during the war, did not take up the slack. The New York committee of safety resolved to underwrite a manufacturing society in 1775, while a cluster of Virginians formed the Manufacturing Society of Williamsburg. Neither society prospered. They limited themselves to promoting textiles and employing the industrious poor, with little to show for their efforts.[41]

The munitions industry suffered from financial distress and a constant drain on manpower. Inflation took a heavy toll, escalating the costs of tools, labor, and raw materials. Although some artisans in crucial occupations held deferments, no consistent policy regarding their long-range utilization emerged. "Exemptions or deferments were parceled out," observed Richard B. Morris, and "effective mobilization of man-power for civilian as well as military tasks was never brought about." Artificers—carpenters, blacksmiths, wheelwrights—were not only in short supply, they bickered over wages among themselves and with their "employers," demanding enough money to keep them ahead of inflation. A New York group once insisted on an immediate pay hike to quadruple their salary.[42] Skilled labor, in fact, remained so scarce and expensive that Congress endeavored to recruit artisans in Europe.[43]

Because munitions could not be produced fast enough to satisfy the insatiable appetite of the military, Congress and the states were bombarded with complaints and pleas for more. George Washington noted the paucity of gunpowder in the army in 1775 and chafed frequently about the quantity and quality of equipment used by his men.[44] John Adams, previously so confident that all would be well, lamented in June 1775 that people "cannot be persuaded to think that it is [as] necessary to prepare for War as it really is. Hence our present Scarcity of Powder."[45] Benjamin Franklin remarked that production increased with time, "yet both arms and ammunition are much wanted."[46] Indeed, some patriot troops had no firearms whatsoever, and several gunsmiths made pikes and spears in lieu of muskets. The situation had apparently improved but slightly in 1776. In March of that year Governor William Livingston of New Jersey warned that his state could not equip its militia unless outside supplies were procured. Pennsylvania had similar troubles, and conditions were as bad if not worse in other states, with New York fearing it would have to disarm its militia if Congress called for new Continental levies. Troops in a few cases waited weeks to take the field because they had no guns.[47]

The sudden need for vast quantities of munitions stretched the fledgling arms industry to the limit, draining reserves as soon as they were gathered. Patriot manufacturers stepped up production, but demand far exceeded supply. Establishing a synchronized, smooth-running industry entailed a sudden shift that a pre-industrial, provincially minded society could not be expected to make without wrenching dislocations. Even if enough muskets could have been produced with the resources available in the colonies, problems inherent with production and distribution would have kept the patriots from being adequately equipped. Furthermore, when patriot troops had guns, they sometimes had no powder. Of the three ingredients used to make gunpowder—charcoal, sulphur, and saltpeter—the first was easy to make, but huge amounts of the second, and only slightly smaller amounts of the third, had to be imported.

Confronted with an inadequate munitions industry at home, Americans were compelled to look to Europe for badly needed supplies. American mills accounted for less than one third of the gunpowder expended during the first two and one half years of the war. The percentage would be even lower if imported cargoes of saltpeter and sulphur were deducted from the total.[48] An overwhelming percentage of the munitions employed during the early war years came from foreign sources. Commercial relations formed before the war strengthened; new associations opened. Those associations became the munitions lifeline of the American war effort, giving Congress and the states time to catch their breath. They also provide an interesting commentary on the character of pre-war technological growth. Shipbuilding during the Colonial Era had flourished and colonial vessels frequented the ports of Europe and the West Indies, so the commercial network of the pre-war years lent itself to the demands of war far more readily than the home-based munitions industry. If not for the tremendous volume of foreign aid brought to the colonies in American-built merchantmen and coasters throughout the war, and from 1775 to 1777 in particular, the patriot war effort might have foundered.

A few munitions cargoes had been whisked in from Europe and the West Indies just before the war. Merchants on Dutch-owned St. Eustatia, soon to become a notorious center of contraband trade, conducted a brisk trade with arms smugglers by late 1774. The Massachusetts committee of safety illegally imported munitions from Bilbao in February 1775.[49] Even a few British merchants could not resist the lucrative profits to be made. General Thomas Gage, while military governor of Massachusetts, cautioned London authorities that numerous colonists sent "to Europe for all kinds of Military Stores."[50] Yet compared with what came in during the war itself, the volume of this trade was inconsequential.

Most committees of safety realized at the outset that importation of munitions from Europe would be required, at least until the home arms industry stabilized. Legislation providing bonuses and loans for powder mills and muskets often had clauses relative to munitions imports. During the summer of 1775 the Maryland convention sent commercial agents to Cape Francois on Hispaniola and to other points in the West Indies. These agents were expected to transship European cargoes to the mainland in small coasting vessels. Plans were also formulated to slip agents into Nantes and Havre de Grace, France, to procure goods directly from Europe. In August 1775 the Pennsylvania committee of safety began seeking munitions in the French and Spanish West Indies. The committee had little choice: while guns made in Pennsylvania averaged twelve dollars apiece, French muskets could be obtained for less than half that amount.[51] New York also sought munitions abroad, and by January 1776 Georgia had followed suit and sent out agents with duties comparable to their Pennsylvania, Maryland, and New York counterparts.[52]

The Continental Congress entered the foreign munitions trade even before it enacted legislation affecting home production. It resolved on July 15, 1775, that each ship transporting war matériel for use by the "continent" be allowed to load and export produce in exchange. Two weeks after the resolution of July 15, Congress set aside $50,000 for a select group of merchants charged with obtaining gunpowder for the Continental army, for which those merchants received a healthy five percent commission.[53] On September 18, 1775, Congress took further steps by creating the Committee of Secret Correspondence. This nine-man committee was empowered to import, with money drawn from the Continental treasury, one million pounds of powder, ten thousand muskets, twenty thousand gunlocks, and forty field pieces.[54] After the treasury reached a dangerously low point, Congress persuaded its contacts to exchange munitions for produce instead of specie. American ship captains accordingly carried cod, lumber, tobacco, and indigo to pay for arms and other military stores.[55]

Congress deployed a network of commercial agents in Europe and the West Indies to facilitate importation, just as many of the states had done. Oliver Pollock, an American residing in New Orleans, funneled supplies through Louisiana into the Carolina backcountry. William Hodge traveled to Europe in May 1776, having been requested by Congress to secure munitions from mercantile firms there. By the close of the year congressional agents were at Santo Domingo, St. Eustatia, Martinique, Curacao, and St. Nicholas Mole in the West Indies, in addition to many of the leading port cities in France and the Dutch Republic.[56]

France, known to be in sympathy with the Americans, was the principal target of congressional agents. Richard Henry Lee expressed the sentiments of many when he told Benjamin Franklin he hoped a formal treaty with that nation would be forthcoming because "we find ourselves endangered by the armaments at present here."[57] Silas Deane arrived in France in May 1776 and began immediate negotiations with arms makers. In December he reported to Congress that he had shipped 80,000 pounds of saltpeter and 200,000 pounds of gunpowder from France via Martinique and 100,000 pounds of powder through Amsterdam. Combining an element of truth with an egotistically dramatic flair, Deane proclaimed that "in some sense the Fate of the United Colonies" depended on the safe arrival of those supplies.[58]

Merchants in France, the Dutch Republic, Spain, Sweden, and Italy viewed the Revolution as an opportunity for expanding their commerce and profits. Though the governments of those countries and their dependencies avoided direct complicity, they seldom interfered with the contraband trade. In a couple of cases they permitted merchants to remove "outmoded" arms from government arsenals for a nominal sum even though their destination was obvious. Quick to see a profit, Dutch arms makers operated their mills at full capacity by mid-1776. They shipped freight bound for the insurgent colonies from Amsterdam or from ports such as Ostend and Nieuport in Belgium. St. Eustatia boomed as an entrepot for illicit trade, as did the French islands of Martinique and Guadaloupe.[59]

Ambitious traders in the Dutch and French West Indies joined dealers on Curacao, Hispaniola, and Spanish Louisiana eager to contact American agents and shipmasters. Don Bernardo de Galvez, governor of Louisiana, countenanced a constant flow of arms, ammunition, cloth and quinine through his province. Some businessmen on British-owned Bermuda and Antigua acted as middlemen in smuggling munitions to the rebels.[60] Thomas Mumford, proprietor of a New Haven, Connecticut, shipping house, was dealing with St. Eustatia merchants by October 1775. Pierre Begozzat, a merchant on Martinique, offered his services to several New England committees of safety early in 1776.[61]

The first substantial contract between a private European company and Congress came in January 1776. Messrs. Penet and Pliarne journeyed to the colonies and visited Washington at his headquarters in Cambridge in the closing days of 1775. Their offer to supply the Americans with a variety of commodities excited congressional interest as well as the backing of Washington. By the following November they had forwarded thousands of dollars worth of military stores and sundry other items to the patriots. In addition to their arrangement with Congress, the Frenchmen opened relations with private firms in New England.[62]

These commercial compacts resulted in a heavy traffic of munitions between Europe, the West Indies, and the rebellious colonies—traffic profitable for both American and European investors.[63] "At most of the Ports east of Boston," reported one loyalist, are "daily arrivals from the West Indies, but most from St. Eustatia; every one brings more or less Gunpowder."[64] The British foreign ministry had reprimanded the Dutch for condoning such practices—apparently with little effect. Major General William Howe complained that the "rebels" received guns and ammunition from Hispaniola, a possession of Spain, a supposedly neutral country. Three months earlier one British gentleman had remonstrated that though the French government gave no overt assistance to the colonists, it did not stop private citizens from doing so. The British admiralty knew this only too well and resentfully advised naval squadrons patrolling the American coast to beware of boats carrying military stores to the colonies flying French colors, holding French papers, and under nominal French masters.[65] To compound the problem, British merchants sometimes continued to trade with the rebels. Thus a British agent in Charleston informed his superiors that gunpowder was smuggled repeatedly into that port disguised as sugar hogsheads from Jamaica.

Shipments of arms and ammunition, whether through the incentive of a merchant or the offices of an American agent, began reaching American ports in significant quantities by mid-1775. In July nearly six and one half tons of powder were unloaded in Philadelphia alone. The *Virginia Gazette* noted that by August upwards of forty tons had been stacked on Philadelphia docks.[66]

Importation mushroomed in the spring and summer of 1776, with merchantmen and privateers putting into Philadelphia, New London, and other ports. One brought in two thousand firearms, over sixty tons of saltpeter, and a large stock of powder; another brought three hundred muskets and over twenty-seven tons of powder. North Carolina merchant John Stanly estimated that in March of 1776 alone 121,000 pounds of powder had been unloaded at various quays in Rhode Island, Connecticut, Massachusetts, New York, and the Carolinas.[67] In May, three unusually large cargoes of powder, saltpeter, and sulphur were delivered at different destinations.[68] By 1777 possibly two million pounds of gunpowder and at least several thousand firearms had been imported, and most of those munitions came from France.[69]

Undeniably the zenith for clandestine aid was reached after Pierre-Augustin Caron de Beaumarchais concocted a scheme to send the Americans illicit assistance. Playwright and author of the *Barber of Seville,* Beaumarchais had been sympathetic to the American cause from the beginning. He realized that influential advisors to King Louis XVI of France, especially the foreign minister, the Comte de Vergennes, hoped to profit by Great Britain's colonial turmoil. France, after all, smarted from losses incurred in the Seven Years War and welcomed an opportunity to see Britain checked. Beaumarchais urged Vergennes and the king to aid the poorly provisioned Americans. Louis XVI granted Beaumarchais's wish that a dummy mercantile firm—Roderique Hortalez et Cie.—be established to assist the patriots. On June 10, 1776, the king authorized the transfer of one million *livres* to a fund with which Beaumarchais could execute his designs. Instead of supplying the patriots with money, the playwright collected firearms from government magazines, firearms that were in mint condition even though being phased out in favor of newer models. Shortly thereafter, Spain, also jealous of Great Britain, matched the French court by donating one million *livres.* With the combined sum, Beaumarchais assembled a truly remarkable stockpile, including 30,000 muskets, 300,000 pounds of gunpowder, and clothes for over 25,000 men.[70]

The first installment of Beaumarchais's aid reached American hands in March 1777. Two of the first three ships leaving France evaded the British. Between them they delivered 21,000 muskets, over 100,000 pounds of powder, and other incidentals.[71] These supplies were quickly distributed among American regiments and, asserted historian Claude Van Tyne, helped make possible the patriot victory at Saratoga.[72]

Beaumarchais's unsolicited relief was unquestionably a tremendous boon. Roderique Hortalez gave a munitions-starved nation a fresh influx of matériel and helped put the American military on a sounder footing. And the guns sent by Beaumarchais, coming as they did from royal arsenals at Maubeuge, Charleville, and St. Etienne, were, generally speaking, of excellent quality. Equally important, they were the largest block of "standardized" weapons brought in during the war. Their uniform design contrasted sharply with the odd mixture of weapons visible in most American regiments, weapons made either under the auspices of committees of safety or smuggled in from Belgian, Dutch, Spanish, and private French arms makers. Yet it should be noted that Beaumarchais's supplies came late in the game. Two years of war passed before they arrived. During those two years the American munitions industry had been far from self-sufficient. Beaumarchais may have been the most important supplier, but he was by no means the first. Roderique Hortalez fit into an already thriving contraband trade.

By 1778 certain alterations had taken place in the munitions carrying trade. Seizures by British warships became all too frequent for European merchants, and they grew reluctant to invest in an increasingly risky business. Beaumarchais in fact lost his entire second fleet that year when it failed to run the gauntlet of British patrols. In addition, the dependency on contraband lessened as France gave more overt support to the Americans, support culminating in the Treaty of Amity and Commerce of May 1778.

Superficially, 1778 also appeared to be a turning point for the home armaments industry. American industry became more proficient and, by increased output, took over part of the market heretofore dominated by foreign firms. There had been a false glimmer of hope that the home industry was finally pushing ahead back in the summer 1776. Robert Morris optimistically informed Silas Deane at that time, "We are better supplied with Powder than formerly . . . our Mills make it fast and some Colonies have had great success in making Salt Petre." There continued to be arms shortages, he conceded, but he believed they would soon disappear. Thomas Paine, writing under the pseudonym "Perseverance," bragged that "cannon we can cast at pleasure, and powder we manufacture faster than we consume it."[73] Judging by reports coming in from the field, Morris and "Perseverance" were considerably off the mark. Yet it did appear that progress was being made and that the patriots were improving their lot by 1777. In March of that year John Adams stated confidently that "there is no longer danger of our wanting arms and ammunition for the future."[74] So it seemed, for the Board of War concluded in August 1778 that munitions on hand were "sufficient."[75] Three years of experience had also brought a semblance of bureaucratic efficiency, with several states, notably Virginia and Pennsylvania, using boards of war to keep up production.[76]

Unfortunately there was another and, as events proved, a gloomier, more realistic side to the picture. In December 1777, just six months before the French officially entered the war, a British agent pleaded with Lord Stormont, the British minister at Versailles, to crack down on the contraband trade flowing out from Europe. "I cannot too often repeat . . . that the very existence of the Rebel Army depends on the arrival of these succours," he wrote, concluding "that if all supplies from Europe are effectively cut-off it is not possible for the Americans to carry on the war."[77] Munitions deficiencies in the American army verified his report, for Continentals as well as militiamen never had enough equipment. Washington wondered with disbelief how this could be true considering how much had been imported.[78] The Board of War's encouraging statement in August 1778 had been preceded by a crisis in April where "we find ourselves at the Eve of the Campaign without Arms fit for . . . even the Continental Troops, without Regard to such as the Militia as may be willing to turn out."[79] In any event, the August 1778 surplus did not last long. The patriots faced shortages in both 1779 and 1780.[80]

As it turned out, supplies from Europe were shipped to the patriots until the end of the war, most of it coming through official channels—namely, the French army and navy—but a small portion still came from private sources.[81] Those who predicted in 1776 and 1778 that munitions shortages were a thing of the past forgot how swiftly a campaign exhausted reserves.

If the home industry had become self-sufficient by late 1778, then Congress would not have encouraged an entire French firm to emigrate and set up shop in the United States. But Congress did just that—convincing evidence that it recognized home manufactures sorely needed assistance. In November 1778 Messrs. Penet and Coulaux sent Congress a memorial proposing that they erect a complete manufacturing complex for cannon, small arms, and ammunition somewhere along the Atlantic Coast. Penet, a longtime friend and provider of contraband since 1775, planned to make everything necessary "so that in a little time the thirteen United States" would not have to depend on "European princes." The Frenchmen would call the factory the "Continental Manufactory of Arms." They wanted to bring over 143 master craftsmen to run the factory and adjacent armory. Seventy-four of the craftsmen would be factory workers; the rest would be foundrymen, blacksmiths, wheelwrights, coopers, carpenters, and nailers. Each European artisan would have several Americans under his supervision so that in time the Americans would learn industrial skills and perhaps erect another factory somewhere else. Penet and Coulaux offered to secure materials for the factory and recruit the workmen at their own expense. Indeed, Penet already had thirty-five German and French gunsmiths repairing firearms in Philadelphia. In return the Frenchmen wanted a grant of land rich with timber for fuel, deposits of iron ore, running water to power their machinery, and a monopoly on government arms contracts.[82]

Congress accepted their terms and in March 1779 ordered the Board of War to draw up a contract "with Mr. Penet & company for 100,000 fire-arms, to be manufactured in these states."[83] Thus the stage was set for the most significant industrial undertaking of the war, an undertaking that could conceivably have pushed American manufacturing technology ahead by several years. Unfortunately, Penet wrote from Nantes that the scheme could not be carried out, for the French government forbade the emigration of skilled workers in time of war.[84] Penet lost a chance to make a small fortune, but American industry by far suffered the greater loss.

It is improbable that the long-suffering American munitions industry was self-sufficient even at the war's end. Adequate stocks may well have existed, but shortages occurred nonetheless. For example, some firearms became unusable after long service and could not be repaired fast enough if at all. Equally as troublesome, countless patriots took their weapons home with them after mustering out, despite orders to leave them with the army. Logistical complications, stemming perhaps from the meager powers of congressional committees like the Board of War or hoarding by state commissaries, still caused problems. The war called for a quantum leap in manufacturing organization and output that the embattled nation simply could not make.

The checkered career of American munitions makers demonstrates how the technology of gunpowder and musket production depended on a long list of variables. Gunsmiths had to step up their output. To do so, they had to recruit more apprentices and master craftsmen, expand their facilities, and gather together stockpiles of the raw materials needed for their firearms. The degree to which they succeeded or failed depended on capital that was hard to secure, the conflicting demands of state and congressional agencies, armies that drained shops of workers, and a society unsettled by fighting that shifted from one scene to another for over eight long years. Patriot merchants involved in importing arms from abroad had fewer such worries. They could rely on an established commercial network and buy wares from European arms makers whose businesses were not subject to the ravages of war.

Munitions imported from Europe and the West Indies were vital to the American war effort. Goods from abroad were sometimes inferior and overpriced by profit-hungry businessmen, yet Americans faced the same predicament with munitions made at home. Although the success of the American move toward independence did not hinge solely on military factors—and John Adams's reflection on the nature of the revolutionary conflict is not to be forgotten—the quest for victory in the field was nonetheless central to the cause. Munitions from Europe enabled the patriots to better equip a poorly armed nation. With provisions from Beaumarchais and others, an expanded domestic industry after 1777, and official French aid in 1778, the most glaring deficiencies were eradicated. But waste, hoarding, and use in battle drained a commissary of military stores that never seemed to have comfortable reserves. Foreign aid, initially clandestine and ultimately open, greased the gears of a creaky American war machine.

What has been said about munitions could easily be applied to textiles. American troops frequently wore clothing imported from Europe, from the war's beginning to its end. The insufficiency of home textiles production, state and congressional concern, and subsequent importation followed a pattern similar to that of the munitions industry. The patriots were simply not prepared for war and could not fend for themselves. They were fortunate that Europeans willingly bailed them out. Pre-war home manufactures agitation had not readied Americans to meet the demands of large-scale, centralized production.

That the states and Congress were hard pressed to procure even basic materials effectively had a debilitating effect on some forms of wartime invention and technological innovation. Even if patriot politicians had been inclined to promote innovation and encourage inventors, they lacked the resources to follow through. Wartime crisis temporarily ruled out any significant advance in manufacturing technology, unless Europeans like Penet and Coulaux could be underwritten by Congress. Americans did well to keep up production by traditional methods whenever they could. To experiment with new industrial processes, given their perilous situation, was almost unthinkable. The United Company of Philadelphia, for example, saw its attempt in early 1776 to introduce new spinning machines "swept aside by the tide of war," and the company itself folded when the British took Philadelphia in 1777.[85]

Yet there remained hope for the future. If nothing else, the war benefited manufactures by awakening Americans to the necessity of improving them. The post-war years consequently saw a rebirth of the home manufactures movement. The munitions industry declined after 1783 because the tools of war were no longer needed, but the emphasis on national mobilization that had sustained the munitions industry lived on. A renewed enthusiasm for improved manufactures was the most enduring legacy of wartime industry.

In the 1760s and early 1770s, manufacturing enthusiasts had decried the existing "evils" of industry—American dependence on imports, the inadequacy of domestic production, insufficient machines. The war made the enthusiasts' point for them: increased production required increased industrial capacity; increased industrial capacity demanded the creation of a more complex technological order. Mobilization without improved production techniques proved inadequate during the war. The advantages of industrialization did not have to be argued to a people who had had to swallow their pride and accept the aid of "papist" France. After the war they were more ready for industrial change than they had ever been before. They followed with greater interest the industrial experiments of post-war manufacturing societies. And those societies were obsessed with bringing the change the public wanted. Just a few months before the collapse of the United Company of Philadelphia in 1777, Robert Jones encouraged its members with his observation that they had laid the "rough solitary stone" of what could become an industrial giant.[86] Jones was right.

NOTES

1. I. Bernard Cohen, "Science and the Revolution," *Technology Review* 47 (1945):367.

2. Meier, "Technological Concept," p. 46; East, *Business Enterprise*, pp. 237-238. Also see the discussion of long-range benefits in J. Franklin Jameson, *The American Revolution Considered as a Social Movement* (Princeton: Princeton University Press; revised ed., 1967), pp. 47-73. Jameson, of course, exaggerated the amount of change actually brought by political independence alone. Even so he understood that the Revolution helped further the cause of change, and to that extent he anticipated later writers who have studied "modernization."

3. East, *Business Enterprise*, pp. 163-164; Victor S. Clark, *History of Manufactures in the United States, 1607-1928*, 3 vols. (Washington, D.C.: Carnegie Institution, 1929), I, 219-220. Curtis Nettles, *The Emergence of a National Economy, 1775-1815* (New York: Holt, Rinehart & Winston, 1962), pp. 23-34, is informative, though Nettles weakened his argument by stating that because of abundant iron and skilled workers, cannon, musket, and hardware shortages "do not figure in the reports of the army's troubles" (p. 43). For a summary of the logistics of the American war effort, see Erna Risch, *Supplying Washington's Army* (Washington, D.C.: Center of Military History, 1981), especially pp. 310-372.

4. *Va. Gazette* (Dixon and Hunter), 1 July 1775, from a letter of 24 December 1774.

5. Adams in *Novanglus*, 6 February 1775, reprinted in John Adams and Jonathan Sewall [Daniel Leonard], *Novanglus and Massachusettenis* (New York: Russell and Russell, 1968), p. 31.

6. PRO/CO, 5/88, fo. 38, as cited in Davies, ed., *Documents*, II (Transcripts), no. 14.

7. Portsmouth in Naval History Division, *Naval Documents of the American Revolution*, 8 vols. (Washington, D.C.: Department of the Navy, 1964-), I, 18-19, 27, 28, 30-31, 37, 38; Massachusetts in a letter from William Bollan to Henry Gibbs of November 1774 in the Massachusetts Papers, 1774-1780; Letters Etc., 1774-1775, no. 27 in the HSP, MSS. copy.

8. *Royal American Magazine* 2 (February 1775):45. The text of the decree is in PRO/CO, 5/29, fo. 36 as cited in Davies, ed., *Documents*, VIII (Transcripts), no. 121.

9. *Pa. Gazette*, 28 December 1774 and 22 February 1775.

10. Willam Moultrie, *Memoirs of the American Revolution*, 2 vols. (New York: New York Times and Arno Press, 1968), I, 63-64, 78. For firearms and gunpowder in general, see M. L. Brown, *Firearms in Colonial America, 1492-1792* (Washington,

D.C.: Smithsonian Institution Press, 1980), which is a good supplement to Harold L. Peterson, *Arms and Armor in Colonial America* (Harrisburg: Stackpole Co., 1956).

11. See Richard B. Morris, "Labor and Mercantilism in the Revolutionary Era," in idem, ed., *The Era of the American Revolution* (New York: Columbia University Press, 1939), pp. 76-139.

12. Hoadly, ed., *Conn. Public Records,* XV, 17-18, 90-91, 199, 287-290.

13. From an act passed by the Connecticut Assembly on 14 June 1776, issued as a broadside printed by Timothy Green in New London. The act concerning inspectors in October 1776 was also printed by Green, as a two-page flyer.

14. Hoadly, ed., *Conn. Public Records,* XV, 251, 325, 329, 372-373; Edmund C. Burnett, ed., *Letters of the Members of the Continental Congress,* 8 vols. (Washington, D.C.: Carnegie Institution, 1921-1936), I, 195; Salisbury in Bishop, *American Manufactures,* I, 511-516.

15. Broadside of 4 November 1775, bearing the signatures of James Warren, speaker of the Massachusetts House of Representatives, and sixteen others, printed by Benjamin Edes in Watertown. Also see the broadside of 30 June and 13 July 1775, by the same printer.

16. Broadside of 8 August 1775, from the New York Provincial Congress; Watkeys in the *Journals of the Provincial Congress, Provincial Convention, Committees of Safety, and Council of Safety of the State of New York,* 2 vols. (Albany: Thurlow Weed, 1842), I, 30.

17. *Resolutions of the Provincial Congress of the Colony of New York for the Encouragement of Manufactories of Gunpowder, Musket Barrels, Musket Locks and Salt* (New York: John Holt, 1776).

18. Force, ed., *Amer. Archives,* 5th series, II, 293.

19. *Pa. Gazette,* 6 December 1775. In October the committee of safety had been willing to take anything it could get. See ibid., 25 October 1775.

20. De Haven in Hazard, ed., *Pa. Col. Records,* X, 482, 506; and idem, ed., *Pa. Archives,* 1st series, IV, 712. County and cannon production in Bishop, *American Manufactures,* I, 553, 572-574; William J. Heller, "The Gunmakers of Old Northampton," *Pennsylvania German Society Publications* 17 (1908):9; and the Papers of the Continental Congress, 1774-1789 (Washington, D.C.: Government Printing Office, 1959, 204 reels microfilm), Item 147, I, 505 (reel 157), hereafter cited as PCC. Also see David L. Salay, "Arming for War: The Production of War Material in Pennsylvania for the American Armies During the Revolution," (Ph.D. dissertation, University of Delaware, 1977); and Salay's article, "The Production of Gunpowder in Pennsylvania During the American Revolution," *Pennsylvania Magazine of History and Biography* 99 (1975):422-442, hereafter *PMHB.*

21. Hening, ed., *Statutes at Large,* IX, 71-73.

22. Force, ed., *Amer. Archives,* 4th series, III, 394.

23. Hening, ed., *Statutes at Large,* IX, 94, 237-238, 305-306; contracts with powder makers in H. R. McIlwaine et al., eds., *Journals of the Council of the State of Virginia,* 4 vols. (Richmond: Division of Purchase and Printing, 1931-1967), IV, 393, 400.

24. Force, ed., *Amer. Archives,* 4th series, IV, 71-73.

25. Ford, ed., *JCC,* II, 190, 218-219; III, 322-323; also see *Several Methods of Making Salt-Petre recommended to the Inhabitants of the United Colonies by the Representatives in Congress* (Philadelphia: William and Thomas Bradford, 1775).

26. Burnett, ed., *Letters*, I, 359; Ford, ed., *JCC*, IV, 169.

27. Burnett, ed., *Letters*, I, 402., Ford, ed., *JCC*, IV, 224.

28. Ford, ed., *JCC*, II, 37, 208-211; VII, 216. For Congress and the logistics of the war economy, see Jack N. Rakove, *The Beginnings of National Politics* (New York: Alfred A. Knopf, 1979), pp. 203, 209-211, 275-280.

29. Inspectors in Ford, ed., *JCC*, V, 713-714; public armorer in Ford, ed., *JCC*, X, 145-146; and the PCC, Item 147, I, 37 (reel 157).

30. PCC, Item 21, 29-31 (reel 30). Springfield in Ford, ed., *JCC*, VII, 266; VIII, 615. Carlisle in Burnett, ed., *Letters*, II, 205, 213.

31. Fred A. Berg, *Encyclopedia of Continental Army Units* (Harrisburg: Stackpole Books, 1972), pp. 11-12; and John C. Fitzpatrick, ed., *The Writings of George Washington*, 39 vols. (Washington, D.C.: Government Printing Office, 1932-1945), X, 277-278.

32. Congress to Washington on 1 September 1777 in the George Washington Papers (Washington, D.C.: Government Printing Office, 1965, 124 reels microfilm), series 4 (reel 43), hereafter GWP.

33. PCC, Item 147, I, 240 (reel 157) and Lancaster County committee of safety minutes, 1774-1777, Peter Force MSS. Collection (Washington, D.C.: Library of Congress microfilm, 1977), Series 8D (reel 48), pp. 130, 132, LC. For deferments and impressments in Virginia of workers and tools, see Hening, ed., *Statutes at Large*, IX, 375-377; X, 397. At the beginning of the war John Fitch (of steamboat fame) reluctantly became a gunsmith in Trenton and was exempted from the draft. See Frank D. Prager, ed., *The Autobiography of John Fitch* (Philadelphia: American Philosophical Society, 1976), p. 55.

34. *The Process for Extracting and Refining Salt-Petre* (Philadelphia: William and Thomas Bradford, 1776); *Essays Upon the Making of Salt-Petre and Gunpowder* (New York: Samuel London, 1776); the *Pa. Gazette*, 19 July 1775 and 7 February 1776; and the *Pennsylvania Packet*, 24 July 1775. Rush is noted in ibid., 24 November 1774, and the *Pa. Journal*, 25 January 1775.

35. Edited by Benjamin West, printed in Norwich, Conn., by Robertson and Trumball.

36. Julian Boyd et al., eds., *The Papers of Thomas Jefferson*, 20 vols. (Princeton: Princeton University Press, 1950-), I, 288.

37. Hazard, ed., *Pa. Archives*, 1st series, IV, 717-718.

38. Massachusetts, Revolution—Military Affairs, Peter Force Transcripts, Series 7E, Box 38, LC, various committee of safety meetings between April and July 1775, most notably the minutes of 8 July 1775.

39. James Warren to John Adams, 13 December 1775, in the "Warren-Adams Letters," *Mass. Hist. Soc. Coll.*, LXXII, 190.

40. Royal Hinman, ed., *A Historical Collection from Official Records, Files, &c., of the Part Sustained by Connecticut During the War of the Revolution* (Hartford: E. Gleason, 1842), p. 273; Burnett, ed., *Letters*, II, 101. As Charles W. Royster has noted, "Americans repeatedly sold defective food, clothing, gunpowder, and other supplies to their own army." See Royster's "The Nature of Treason: Revolutionary Virtue and American Reactions to Benedict Arnold," *WMQ*, 3rd series, 36 (1979):177.

41. Force, ed., *Amer. Archives*, 4th series, III, 1424-1426; *Va. Gazette* (Purdie), 23 August 1776.

42. Morris, *Government and Labor*, pp. 280, 297, for a general statement; plus see the letters from Udny Hay to Nathanael Greene of 9 January and 2 October 1779 in the Nathanael Greene Papers, VIII, 42-42a and III, 91, respectively, APS MSS., for the New York artificers.

43. Ford, ed., *JCC*, IX, 883-884.

44. Force, ed., *Amer. Archives*, 5th series, I, 107; Fitzpatrick, ed., *Writings of Washington*, IV, 288-289, 326. Also see the letter of John Sullivan to John Langdon of 4 September 1775 in the Force Transcripts, Series 7E, Box 8, Alfred Elwyn Papers, LC.

45. Burnett, ed., *Letters*, I, 118.

46. Force, ed., *Amer. Archives*, 4th series, IV, 353-354.

47. Livingston in the PCC, Item 68, 109 (reel 82); Pennsylvania in Hazard, ed., *Pa. Archives*, 4th series, III, 610-611; New York in Force, ed., *Amer. Archives*, 4th series, III, 914; and Paul H. Smith, ed., *Letters of Delegates of the Continental Congress*, 9 vols. (Washington, D.C.: Library of Congress, 1976-), I, 678-679. Also see Maryland in Force, ed., *Amer. Archives*, 4th series, IV, 911-912.

48. Arthur P. Van Gelder and Hugo Schlatter, *History of the Explosives Industry in America* (New York: New York Times and Arno Press, 1972), pp. 38-68, argued that American mills produced about 33 percent for the first two and one half years. Orlando W. Stephenson, "The Supply of Gunpowder in 1776," *American Historical Review* 30 (1925):277 gives lower figures; also see idem, "The Supplies of the American Revolutionary Army," (Ph.D. dissertation, University of Michigan, 1919).

49. See the Massachusetts Papers, 1774-1780; Letters Etc., 1774-1775, no. 31, HSP MSS. copy. Also see the letter from Elbridge Gerry of 2 July 1775 in ibid., Massachusetts Board of War, I, no. 57.

50. Carter, ed., *Correspondence of Gage*, I, 385-386; St. Eustatia in the PRO/CO, 5/138, fos. 392-397d, as cited in Davies, ed., *Documents*, VII (Calender), nos. 559-559ii.

51. Maryland in Force, ed., *Amer. Archives*, 4th series III, 496-497; price of guns in Pennsylvania in Henry S. Commager and Richard Morris, eds., *The Spirit of Seventy-Six* (New York: Harper and Row, 1967), p. 775.

52. *Naval Documents*, III, 582-583; Burnett, ed., *Letters*, I, 165, 221; *Journals of the Provincial Congress*, I, 129.

53. Ford, ed., *JCC*, VII, 84-86, 184-185, 210; III, 280, 306-308.

54. Ibid., II, 253-256; III, 336. For an overview see Elizabeth M. Nuxoll, "Congress and the Munitions Merchants: The Secret Committee of Trade During the American Revolution, 1775-1777," (Ph.D. dissertation, City University of New York, 1979).

55. Exports in Ford, ed., *JCC*, III, 308.

56. Pollock in the PCC, Item 50 (reel 64); and letter of 30 May 1776 from Henry Laurens in the Henry Laurens Papers, I, HSP MSS.; and PCC, Item 37, 1-3 (reel 44) for William Hodge. For other agents, see Francis Wharton, ed., The *Revolutionary Diplomatic Correspondence of the United States*, 6 vols. (Washington, D.C.: Government Printing Office, 1889), I, 129-131. One of the key agents was William Bingham, who was stationed in Martinique. See the secret committee's directive to him of 3 June 1776 in the Simon Gratz Autograph Collection, Case 1, Box 19, HSP MSS.

57. James C. Ballagh, ed., *The Letters of Richard Henry Lee*, 2 vols. (New York: Da Capo Press, 1970), I, 218.

58. Deane memorial of 22 August 1776 in Stevens, ed., *Facsimiles*, VI, 580.

59. Friedrich Edler, *The Dutch Republic and the American Resolution* (Baltimore: Johns Hopkins University Press, 1911), pp. 41, 43; J. Franklin Jameson, "St. Eustatius in the American Revolution," *American Historical Review* 8 (1903):683-700; and H. A. Barton, "Sweden and the War of American Independence," *WMQ*, 3rd series, 23 (1966):410-411. Denmark was one of the few to refuse to sell arms to the patriots; see Force, ed., *Amer. Archives*, 4th series, III, 942.

60. Francisco Morales Padron, *Spanish Help in American Independence* (Madrid: Publicaciones Espanolas, 1952), pp. 16-39; and PRO/CO, 5/138, fos. 154-161d, cited in Davies, ed., *Documents*, X (Calender), no. 452i. William B. Kerr, *Bermuda and the American Revolution* (Princeton: Princeton University Press, 1936), pp. 47-55; *Naval Documents*, I, 1169-1170, 1188; and Antigua in PRO/CO, 5/122, fos. 306-315d, as cited in Davies, ed., *Documents*, X (Calender), nos. 629i-629vii.

61. *Naval Documents*, II, 521; III, 1297; IV, 40-41.

62. Ford, ed., *JCC*, III, 465-467; Fitzpatrick, ed., *Writings of Washington*, IV, 159-163; *Naval Documents*, III, 693; VI, 476n. A promise to keep Washington supplied with munitions is in a letter of 3 August 1776 in the GWP, series 4 (reel 37).

63. East, *Business Enterprise*, pp. 36-43.

64. *Naval Documents*, III, 710.

65. Ibid., I, 949; III, 254; Stevens, ed., *Facsimiles*, XIII, 1302-1304, for the issue of French complicity and PRO/CO, 5/139, fos. 34-37d, as cited in Davies, ed., *Documents*, X (Calender), no. 1208i.

66 Ford, ed., *JCC*, II, 204; *Va. Gazette* (Purdie), 18 August 1775.

67. *Naval Documents*, IV, 164, 219; Boyd, ed., *Papers of Jefferson*, I, 285; John Stanly in the *Va. Gazette* (Dixon and Hunter),6 April 1776.

68. *Pa. Gazette*, 20 May, 12 June and 3 July 1776.

69. Stephenson, "Supply of Gunpowder in 1776," pp. 271-281, estimated the total at 1,454,000 pounds (plus about 478,000 pounds of saltpeter). I have revised that figure upward, for the simple reason that Stephenson relied almost exclusively on official records and apparently did not consult newspapers. Six large shipments from May to August 1776 alone reported in the papers accounted for over 150,000 pounds. A spot check of Stephenson's colony-by-colony breakdown led me to conclude that such shipments did not always make it into official records. Also, Stephenson assumed incorrectly that the patriots were not very dependent on Europe for firearms (p. 277n).

70. Beaumarchais in Wharton, ed., *Diplomatic Correspondence*, I, 369-371, 453-455; II, 129-131, 148, 262; Elizabeth S. Kite, "French 'Secret Aid' Precursor to the French-American Alliance, 1776-1777," *French American Review* 1 (1948):143-152; and James B. Perkins, *France in the American Revolution* (Boston: Houghton Mifflin Co., 1911), pp. 92-93, 106-107. Also see Richard Van Alstyne, *Empire and Independence* (New York: John Wiley and Sons, 1965), pp. 84-86; and Helen Augur, *The Secret War of Independence* (New York: Sloan and Pearce, 1955), pp. 21-24, 80-98, for clandestine aid.

71. PCC, Item 156, 478-482 (reel 176), a breakdown of the cargoes by Tronson du Coudray. Beaumarchais sent a total of eight ships in 1776–1777.

72. Claude Van Tyne, "French Aid Before the Alliance of 1778," *American Historical Review* 31 (1925):40.

73. *Naval Documents*, V, 384; Force, ed., *Amer. Archives*, 5th series, III, 1394.

74. Burnett, ed., *Letters*, II, 313.

75. PCC, Item 147, II, 193 (reel 157).

76. For the Virginia Board of War see Hening, ed., *Statutes at Large*, X, 17-18, 198-199; for Pennsylvania see the Gratz Collection, Pennsylvania Series, Board of War, Case 1, Box 18, HSP MSS., under Thomas Wharton. Massachusetts had had a Board of War from the beginning.

77. Stevens, ed., *Facsimiles*, XXI, 1822. In 1778 the Board of War decided it was cheaper to buy steel from France than make it at the Andover Ironworks in New Jersey. See the PCC, Item 147, II, 47 (reel 157).

78. Militia shortages in 1777 and 1779 are noted in the PCC, Item 69, I, 363 (reel 83) and Item 70, 327 (reel 84). Also see Fitzpatrick, ed., *Writings of Washington*, VIII, 368-369; and Hazard, ed., *Pa. Archives*, 1st series, VIII, 434, 442, for 1780 shortages.

79. PCC, Item 147, II, 13-14 (reel 157).

80. To take Virginia as an example, arms shortages among militia units in 1781 were worse than they had been in 1779, when stocks were quite low. See the letter of Robert Forsyth to Nathanael Greene on 19 May 1779 in the Nathanael Greene Papers, V, 33, APS MSS.; and George Weedon to Baron von Steuben on 1 April 1781 and to Lafayette on 16 June 1781 in the George Weedon Military Correspondence, 1777-1786, pp. 67 and 133, respectively, in the APS MSS.

81. For 1780 shipments from France and the West Indies, see the bills of lading in the Henry Knox Papers (Boston: Massachusetts Historical Society, 1960, 55 reels microfilm), V, 104-105 (reel 5); and the dependency of the army on the French in Burnett, ed., *Letters*, V, 318, 447, 576, 578.

82. PCC, Item 41, VIII, 60-63 (reel 51), and X, 293-299 (reel 52).

83. Details of the contract are in ibid., Item 147, III, 117-118 (reel 158), and Item 19, V, 73 (reel 26); also see Ford, ed., *JCC*, XIII, 16-17, 303-304; XIX, 241.

84. PCC, Item 78, XVIII, 291-293 (reel 100).

85. Jeremy, "British Textile Transmission to the U.S.," p. 29.

86. Robert B. Jones, "An oration delivered in the college of Philadelphia before the united company of Philadelphia for promoting American manufactures," *American Museum* 5 (March 1789):265-267.

4 GENIUS FRUSTRATED: THE VISIONARIES

There were hundreds of inventive craftsmen scattered around the colonies in 1775. Nevertheless, despite the size of this pool of ingenuity, inventors who designed unusual or unsolicited inventions had to fight for recognition, and those few who came forward with revolutionary new devices that they thought would help win the war usually went away disappointed. The more radical the invention, the less likely were the chances of its being used. Because American society was in no position to experiment at will with strange new devices, the War of Independence closed opportunities for invention in some fields even as it opened opportunities in others.

Brooke Hindle proffered one explanation for the lack of identifiable scientific activity during the war, and his explanation could be applied to some forms of invention just as easily. In brief, he argued that the leading scientists of the pre-war decades were distracted by other duties.[1] What was true of scientists tended to be true of many inventors. Thresher inventor John Hobday of Virginia managed a salt works. Mechanical wizard Christopher Colles scrimped for a living, conducting road surveys when not instructing artillery officers in their art.[2] Franklin sat as a member of Congress and the Pennsylvania committee of safety and, after mid-1776, attended to diplomatic affairs in France. David Rittenhouse was likewise preoccupied, serving on the Pennsylvania committee of safety and also as a representative in the state assembly. Thomas Jefferson dealt with pressing business, first in Philadelphia and later as governor of Virginia. Perhaps he captured the sentiments of Colles and the others when he lamented that men of genius should be so distracted. "The world has but one Rittenhouse," he complained; affairs of state ought to be left to more "ordinary men."[3]

Franklin, Rittenhouse, Jefferson, and most of their colleagues in the scientific community were, for all intents and purposes, temporarily removed from that loose association. With their members drawn off into other fields, colonial scientific societies atrophied. The Virginia Society for Advancing Useful Knowledge met only infrequently from 1775 to 1777 despite the plea of its president, John Page, that such societies should be most productive during troubled times.[4] The American Philosophical Society, the one central clearinghouse for inventors in the prewar years, was all but disbanded from 1776 to 1780.

American society was not organized, intellectually or materially, to capitalize on the potential of many inventions. The handful of scientific societies springing up in the 1770s supplied needed patronage but were not by themselves enough to sustain invention nationwide. When they ceased to meet, inventors had no alternative forum for their ideas. The new national government could not provide the type of assistance most inventors wanted. The power of the Continental Congress was circumscribed to begin with; sectional jealousies, combined with personal animosities, worsened an inventor's chances of obtaining support. Enmity and prejudice were so strong, grumbled John Adams, that it "is almost impossible to move anything" through Congress.[5] Though Adams exaggerated, an inventor of limited means and lacking influential friends faced certain disadvantages, especially since committeemen assigned to review petitions were rarely if ever selected for their expertise in technical subjects.

Congressional inadequacies, however, were symptomatic of a larger social phenomenon. In April 1777 Congress requested the states to keep it informed of "all discoveries and improvements in the arts of war," yet no state made much of an effort to do so.[6] The states, inexperienced in such matters, quite possibly did not know how to execute Congress's request because they had not been that directly involved in promoting invention. What had been true during the colonial period often held true through the war: inventions of a dramatic nature requiring substantial change were not often given support. Some fairly modest inventions were practically ignored as well. Oziel Wilkinson of Rhode Island developed a new way to cut nails, and Oliver Evans of Delaware invented a new machine for cutting wire teeth for cards, but neither man's genius was exploited during the war.[7] Similarly, Thomas Paine planned experiments on a steel crossbow capable of shooting fire arrows the width of the Delaware River, but knowledge of his scheme did not go beyond a small circle of friends, and it seems doubtful that he ever conducted the experiments.[8]

Only a handful of inventors approached Congress directly—or indirectly through George Washington—from 1775 to 1781, and they ended up with little to show for their efforts. Others concentrated on their respective state governments, in at least one instance with considerable success, as will be shown in the next chapter on the Delaware River defense system set up by Pennsylvania. Perhaps some inventors were reluctant to petition Congress, suspecting it either could or would do nothing for them. The possible benefits accruing from public exposure may not have occurred to many inventors. Some who stepped forward had nothing worthwhile to offer.

The first inventor to make overtures to Congress was Captain John Macpherson, and he was one of those who, alas, did not have much to offer. Macpherson is significant, not because his ideas were revolutionary, but because he was a peculiar character. Some members of Congress who dealt with Macpherson may have formed their opinions about the value of wartime invention, if not invention in general, in response to their experience with him. And Macpherson was proof enough that inventors often gained a hearing because of their salemanship and social connections rather than by the merit of their ideas.

Macpherson was born in Edinburgh, Scotland, in 1726. Following a strange childhood running the gamut from criminal pranks to grave robbing for his pharmacist uncle, all of which he wrote about with gusto later in life, Macpherson stole away, went to sea, and eventually emigrated to the colonies. He became wealthy after successful exploits as a privateer in the wars for empire and in 1764 was made a burgess of Edinburgh in recognition of his service. He made his home on an elegant estate northwest of Philadelphia known as Mount Pleasant. Macpherson married well and sired four offspring. His two sons attended the College of New Jersey. John, the youngest, became an officer in the Continental army and was slain alongside Richard Montgomery at Quebec. William, the eldest, rose to prominence in Pennsylvania politics after the war.[9]

Macpherson suffered from a form of dementia, a neurological disorder which brought on seizures and temporary loss of reason. In 1769 his family had him confined in a secluded hut on the estate for a time, a "betrayal" for which he never forgave them. Plagued by a persecution complex, he swore that he was victimized by a cabal, an "evil Triumvirate" headed by former friend John Dickinson. In a series of vituperative pamphlets Macpherson charged Dickinson and several others with conspiring against him and contriving his confinement because he knew the "truth" about Dickinson's *Letters from a Farmer in Pennsylvania*.[10] Macpherson never divulged the exact nature of that "truth" (because he wanted the issue decided in court), but he implied that the letters were not really the product of Dickinson's genius—at least in the manner most readers believed. Not surprisingly, many of Macpherson's acquaintances regarded him as an eccentric, and rightly so.

Macpherson was, among other things, an inventor. In 1771 he applied unsuccessfully to the Pennsylvania Assembly for some sort of reward for his "grain grinder" and a special machine for raising water.[11] When the war came, he tried to interest influential out-of-state congressmen with still other ideas. He went first to John Adams. Macpherson shared with Adams a secret plan that he was "sanguine, confident, positive" would make possible the destruction of "every man-of-war in America." He invited the impressionable Adams to join him for dinner a week later, where he again hinted at the "secret."[12] Finding Adams receptive, Macpherson pressed for a larger audience. He notified Congress that he would disclose to a chosen few his plan to destroy the British squadron in Boston harbor. On October 16 a special three-man committee was appointed to converse with him. They, like Adams, went away impressed and recommended that Macpherson "ought to repair" to the camp at Cambridge. Congress sent him forward with its blessing and $300 to cover expenses. John Hancock's note to George Washington introducing the inventor explained that the plan was very secret, having as yet only been divulged to the select committee. "These Gentlemen reported, that the scheme in theory appeared practicable, and though its success could not be relied on without experiment, they thought it well worth attempting."[13] Macpherson also carried a note to James Warren written by Adams, characterizing him as a "Genius." Caught up in the excitement of Macpherson's mysterious mission, Adams added, "Ask no Questions and I will tell you no false News."[14]

Macpherson wasted no time. He arrived in Boston on October 27, although to a much different reception than he had anticipated. Washington and his staff listened patiently to the Pennsylvanian but found his plan wanting. Judging by the tone of a letter Washington subsequently sent to Congress, he was somewhat miffed that Macpherson had been dispatched with such fanfare (or as Charles Francis Adams noted, the general made a "sly hit" on Macpherson's scheme in a rare sarcastic moment). Washington, his chief engineer Richard Gridley, Gridley's assistant, Henry Knox, and a third officer, probably Rufus Putnam, decided that Macpherson's plan would prove "abortive." All Macpherson had intended to do was build a fleet of small row galleys armed with a single cannon each. These galleys would supposedly be superior to large warships in narrow bodies of water like Boston harbor. Macpherson, after receiving Washington's adverse report, nonetheless desired to stay on and assemble his mini-fleet, at his own expense if need be. The commander in chief chose to wash his hands of the affair and persuaded the downcast inventor to go home.[15]

Macpherson's idea of using galleys was not that ill-conceived, but it hardly required such secrecy. Galleys were in fact used successfully on Lake Champlain, the Delaware River, and Long Island Sound later in the war. Champions of the little crafts included Benedict Arnold, Thomas Paine, and Josiah Quincy.[16] Nevertheless, Macpherson's particular plan had numerous technical drawbacks that Washington and his engineers saw immediately. They were perhaps disappointed Congress had not.

If the Macpherson episode had ended here, it probably would not have had any lasting effect. But Macpherson refused to give up. Congress heard from him again in March 1776, when he declared that he was more convinced than ever that his inventions—he had several by then—were extremely important to the country. He offered to disclose details to a special investigatory committee. Receiving no response, Macpherson petitioned Congress again, guaranteeing that he could sweep the British navy from American waters with a dozen of his specially designed galleys. He might even outfit two complete boats at his own expense, staking his reputation on his ability to sink "an english man of war in very little time." Congress did not take the bait. To its amazement, it next heard Macpherson claim that the committee that first interviewed him had promised him the post of commander in chief of the American navy. A special committee assembled in July 1776 found, as most suspected, that Macpherson had no written evidence to support his claim. The surviving members of the original committee adamantly denied making any such promise. Therefore Congress dismissed Macpherson's application as "unreasonable."[17]

Undaunted, the ex-privateer continued to haunt the halls of Congress. He periodically asked for aid or served notice that he had yet another invention. In March 1778 he asked Congress for a loan of £1000 for a trip to France, where he proposed to build a special high seas cruiser to prey on British merchantmen. He was refused. In a couple of cases Congress granted Macpherson's request for powder to conduct "experiments," perhaps to keep him busy and out of their way.[18] Macpherson never relented, fruitlessly seeking congressional subsidies long after the war.

Even after Macpherson's instability became fairly obvious to members of Congress, special committees were nonetheless assigned to examine his schemes. Nothing he submitted was turned down out of hand. Of course this was due in part to Macpherson's social standing. He customarily moved in important circles and commanded a certain amount of attention because of it. And Macpherson proved to be an accomplished salesman, as evidenced by the enthusiastic if naive response he elicited from John Adams and others. Congress willingly granted a hearing, but a hearing by no means indicated support would be forthcoming or, for that matter, was even being considered. Macpherson, while a pathetic figure, was also a nuisance and may have left a bitter taste in too many mouths. Although Washington, Adams, and the rest remained silent on the subject, Macpherson may have made acceptance all the more difficult for other inventors.

David Bushnell, the submariner of the Revolution, should therefore be considered in connection with John Macpherson. Where Macpherson was wealthy and well-connected, Bushnell was poor and had few friends in high places. More important, Macpherson was a relentless salesman and promoter of his own projects, whereas Bushnell was reticent and reluctant to push. Hence the irony connecting these two men. Macpherson, by the force of his personality, gained the ear of Congress; the much less aggressive Bushnell apparently did not seriously try. Indeed, Bushnell had a "passion for anonymity."[19] The one inventor whose ideas were truly revolutionary—and practical—would remain obscure, all but forgotten by his own generation. Bushnell needs to be examined in some detail, both because of the innovativeness of his ideas and because, more than any other wartime inventor, he was a victim of circumstance, a personification of the frustrated genius.

David Bushnell was born in 1740, the son of a modest farmer in Pochaug, the west parish of Saybrook, Connecticut. Ships, the sea, and mechanics interested young David more than farming, so the thought of following in his father's footsteps did not excite him. He tinkered about on the family farm, hoping someday to move on. He did not fulfill his ambition of attending Yale College until he was thirty-one. Bushnell departed for New Haven and school in 1771, poor, unmarried, and without many prospects. He soon became popular among his fellow students, despite, or perhaps because of, his unusual age for a freshman. Furthermore, he endeared himself to several tutors and numbered among his friends Timothy Dwight, who later eulogized him in *Greenfield Hill*.[20]

Bushnell spent a good deal of his time at Yale in the laboratory, where he seemed to be most at home. Late in his freshman year he conducted experiments which verified the findings of Sir Edmund Halley that powder submerged in an airtight compartment under water could be exploded. By his senior year Bushnell had progressed from experiments with powder to devising a "torpedo" for holding an explosive charge. The torpedo was an ingenious egg-shaped contrivance of two hollowed-out pieces of wood filled with powder and joined by iron bands. Bushnell inserted a clock mechanism that tripped a hammer, flint, and steel from the lock of an ordinary musket to ignite the powder. Nearing graduation in the spring of 1775, he toyed with the idea of making a submersible vessel with which to deliver the torpedo.[21]

Bushnell probably began working on his submarine in April or May 1775, when classes were temporarily suspended and graduation postponed because of the troubles at Boston. He worked in seclusion in a shed on the Connecticut River, near his hometown of Saybrook. Once completed Bushnell's submarine, dubbed the "American Turtle," was an odd-looking device. Bushnell later described it as bearing a resemblance to "two upper tortoise shells of equal size, joined together." The six-inch-thick oak hull, bound with iron and caulked with tar, had an aperture at the top, through which the pilot entered. The interior of the submarine was barely large enough for the pilot and a thirty-minute supply of air. The top was crowned with a brass conning tower containing three watertight portholes, replete with tiny glass windows. The inside

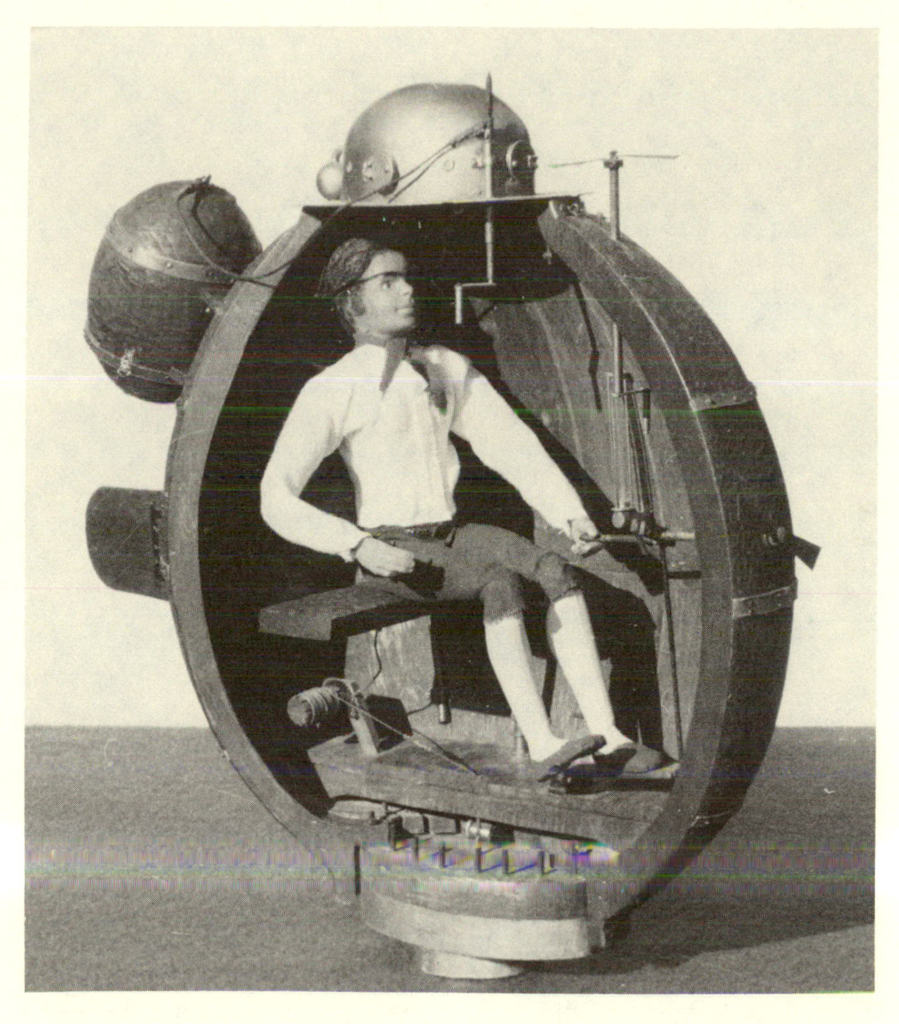

8. Model of David Bushnell's "Turtle." This cutaway view demonstrates the deceptive simplicity of Bushnell's machine. Courtesy of Robert M. Speck.

of the "Turtle" was cramped, with just enough space for a compass and depth gauge. The pilot controlled the rudder by a hand tiller and two hand cranks connected to paddles that moved the boat vertically and horizontally at about three knots. Bushnell's submarine also had a brass valve to allow water in and a brass pump to force water out for surfacing. Over two hundred pounds of lead, much of it detachable, was fixed to the "Turtle's" bottom as additional weight. The machine could dive or

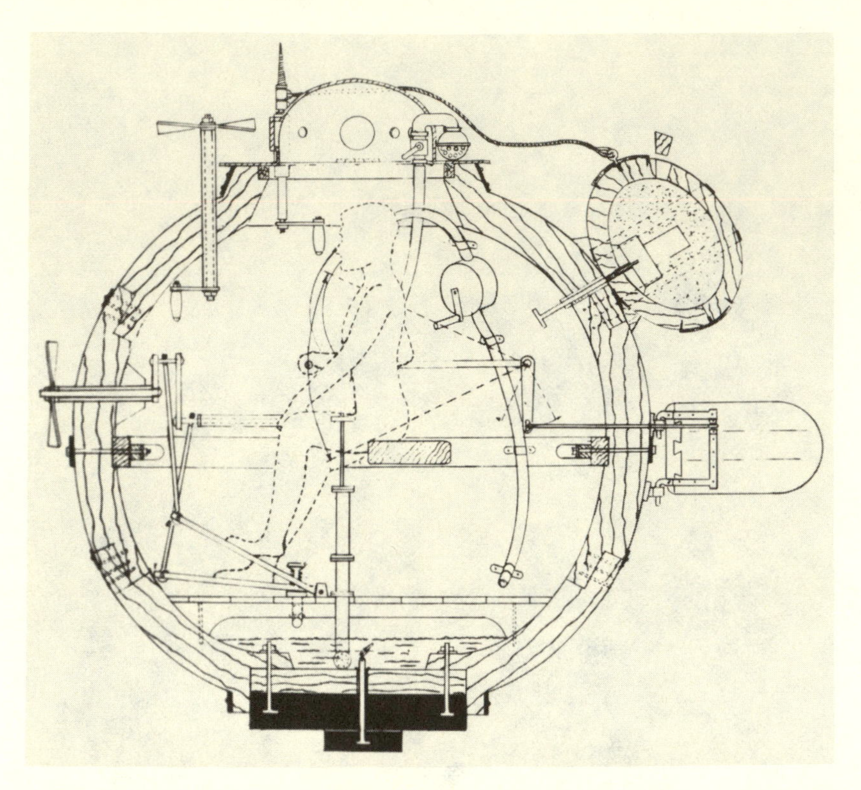

9. Schematic drawing of David Bushnell's submarine. This blueprintlike representation of the "Turtle" accentuates Bushnell's sophisticated design and efficient, economic use of space and materials. Compare this drawing with the model in Fig. 8. Drawing by Frank Tinsley on deposit at Submarine Force Library, Groton, Conn., amended by Robert M. Speck. Courtesy of Robert M. Speck.

submerge quite handily, and two self-sealing brass ventilation pipes passed through the crown for circulating air on surface runs.

Bushnell also equipped his submarine with an iron screw worked from the inside, designed for drilling into the hull of a wooden warship. He attached a line leading from a watertight, one hundred and fifty pound charge resting on the back of the "Turtle" to the screw. After seating the screw tightly in his target, the "Turtle's" pilot could release it and the powder charge and paddle away to a safe distance. Ignited by the firing mechanism Bushnell developed as an undergraduate at Yale, the charge would explode, the resulting concussion bursting the target ship's seams.[22]

Although short of money, Bushnell financed the work himself, not wanting to draw any notice until he could be sure of success. Besides, the financial assistance offered by the state of Connecticut at this point "was so Inconsiderable he refused it."[23] Only Bushnell's brother Ezra and a local ironmonger commissioned to make metal parts for the "Turtle" knew the details of his work in the early stages. As the machine neared completion, the young inventor also gave long-time friend Dr. Benjamin Gale a look at his work, but supposedly no one else was privy to the details. Still, Bushnell did not have anything resembling tight security, so news of his scheme leaked out. John Lewis, a Yale tutor, mentioned the project to Ezra Stiles, and Stiles in turn undoubtedly passed the information along.[24] Silas Deane, then serving in Congress as a Connecticut delegate, wrote Dr. Gale and inquired about rumors he had heard "of a certain new invention for destroying ships."[25] He requested verification, which Gale gladly furnished (though it would seem likely that the good doctor checked with Bushnell first). He sent Deane a brief description of the "Turtle" and closed with the admonition that if Bushnell succeeded, "a stipend for life, and if he fails, a reasonable compensation for time and expense is his due."[26]

Three days before he received Deane's letter, Gale had sent off a note to Benjamin Franklin at Bushnell's request, describing the submarine and its successful trial runs in full. Franklin subsequently wrote Deane, "I shall be curious to hear some more Particulars of your new mechanical Genius."[27] By October knowledge of Bushnell's activity had even reached Samuel Osgood in Boston, who informed John Adams in Philadelphia that "the Water Machine from Connecticut is every Day expected in Camp."[28]

Thus at least three prominent members of Congress knew of Bushnell by the fall of 1775. Yet Adams, Deane, and Franklin did nothing to bring this news to the formal attention of Congress. They were curious, Franklin in fact stopping at Saybrook to visit Bushnell on his way to Boston in October, but apparently little else. This would not seem particularly strange, unless Bushnell's case is compared with Macpherson's. Macpherson garnered substantial support with only slight difficulty. Key members of Congress knew of Bushnell's "Turtle" at least a full month before Macpherson met with John Adams and hinted at his plan, but they chose not to champion Bushnell's cause.

This seeming inconsistency can be explained in a number of ways. Bushnell was comparatively young and virtually unknown in congressional circles. Though older and more experienced, Macpherson, like Bushnell, had to cultivate his connections. Macpherson had the double advantage of wealth and proximity. His nearness to Philadelphia and the ear of Congress, coupled with his material resources, enabled him to wine and dine the right people. Moreover, Bushnell exacerbated his natural disadvantages by his own reticence. His policy of remaining silent until his theories were proven was wise, wiser in the long run than Macpherson's bombastic approach; but once sure of himself, he needed to be more aggressive. Instead he relied on his invention to sell itself. A final and no less important factor was that the very ingenuity of Bushnell's scheme may have worked against him. Macpherson's proposal for row galleys was much easier to comprehend than Bushnell's because the "Turtle" was a fairly sophisticated device. Macpherson only called for modifications in a common boat design, with tactics adapted to those modifications. Bushnell proposed a different type of warfare altogether.

If Bushnell had been able to transport the "Turtle" to Boston before the winter of 1775-1776, he might have gotten the public exposure he needed. Unfortunately, he was delayed because he ran out of money and had to stop working on the "Turtle" in September. By early November he had scraped together some funding, put on the finishing touches, and prepared to rush to Boston. Then he ran into technical problems. A poorly constructed ballast pump had to be returned to the ironmonger doing the metal work for the "Turtle." There was also the problem of illumination. Bushnell wanted to launch the submarine at night, so he needed artificial light for the depth gauge. He eventually coated the compass dial and cork float inside his depth gauge with phosphorous, but for the time being he was stumped. Because of these complications, he kept the "Turtle" in storage over the winter. Bushnell had to wait until the following spring to prove the effectiveness of his invention.[29]

He did not remain idle, however. He finally applied to the Connecticut council of safety in February 1776 for assistance in paying his bills. The council listened sympathetically and generously voted him £60 to defray costs.[30] Thus unencumbered of burdensome expenses, Bushnell looked ahead to the spring. Governor Jonathan Trumbull, a staunch supporter of Bushnell who had recommended the course taken by the council, made no formal communication to Congress of Bushnell's progress. As before, news of the "Turtle" spread only by word of mouth.

By the summer of 1776 Bushnell was most anxious to test his submarine and sailed on a borrowed schooner with the "Turtle" in tow down Long Island Sound to New York. David's brother Ezra had been stricken with a fever and could no longer continue as pilot, a crushing blow to Bushnell's hopes. Ezra had been the pilot since the first successful trial run months earlier. His expertise would be sorely missed, for David was too frail to operate the machine for any length of time himself. The budding inventor had to generate interest among American officers stationed in New York to find someone capable of navigating the "Turtle." He managed to do so. Sergeant Ezra Lee and two other volunteers for hazardous duty aboard fireships were selected to work with Bushnell. They sailed back up to Saybrook, where Lee and the others received a crash course in piloting a submarine.

In the meantime, Long Island fell to the British, and Washington withdrew his shattered army to Manhattan. But Bushnell had not been forgotten. One officer wrote to Joseph Trumbull, son of the Connecticut governor, "Where is Bushnell, and why dont he attempt something."[31] Finally satisfied that Lee and the alternate pilots were ready, Bushnell slipped the "Turtle" back down Long Island Sound to New Rochelle and carried it overland to the Hudson. Samuel Parsons, Ezra Lee's brother-in-law, supervised the move. He had helped Bushnell previously, confident that "though the event is uncertain, the experiment under our present circumstances is certainly worth trying."[32]

Bushnell settled on HMS *Eagle*, flagship of Lord Richard Howe, as his target. In the wee hours before dawn on September 6, boatmen towed the "Turtle," with Lee inside, out from Manhattan and set it adrift just above the British warship. Lee tried desperately to maneuver his craft under the warship's keel. He was not adept enough to accomplish his mission, or, as he stated later, he could not pierce the *Eagle's* hull because of iron rods at the screw's point of entry. Lee pushed off, fearing daylight might disclose his presence. He cut the torpedo loose and cranked furiously for safety. Half an hour later the torpedo, a long way from its target, exploded violently, throwing "into the air a prodigious column of water, resembling a great water spout, attended with a report like thunder."[33]

Bushnell's golden opportunity passed, not because of any defect in design or craftsmanship, but because of poor luck. Had Lee penetrated the *Eagle's* hull, or had Ezra Bushnell, a far more accomplished pilot, not been ill, the results might have been different. If the "Turtle" had been successful, Bushnell would have become an instant hero because the Americans had little to cheer about at that time.

A small gathering of notables witnessed the event. George Washington, one of those present, wrote to Thomas Jefferson several years after the war and recalled his impressions. He referred to Bushnell as a "man of great Mechanical powers . . . fertile of imagination . . . and a master of execution."[34] Nonetheless, Washington lacked faith in the practicality of the submarine. He argued that its "novelty," the difficulty of underwater navigation, and the uncertainty of hitting the target militated against success. Washington viewed the "Turtle" as a curiosity and, though willing to give Bushnell a chance to try his luck, did not really see any potential worth to the invention. No official word of the episode was communicated to Congress, and mention of Bushnell's near miss only survived in a handful of private journals and correspondences.

After the abortive attack on the *Eagle*, Bushnell dejectedly shifted the "Turtle" to a protected inlet above Fort Washington. The ill-fated inventor's hopes were delivered a mortal blow a month later when a British naval sortie ran past the guns of the fort and sank the schooner carrying his submarine. Bushnell recovered his damaged machine, but he never used it again, noting that "the situation of public affairs was such, that I despaired of obtaining the public attention, and the assistance necessary."[35] The vessel's ultimate fate is shrouded in mystery. Perhaps Bushnell took it back to Saybrook and later scuttled it to prevent its falling into British hands, or perhaps it suffered irreparable damage when sunk. In any event the patriots lost one of the few inventions of any note to be brought forward during the war.

Though abandoning further work on the "Turtle," Bushnell did not give up experiments with his underwater explosives. The submarine, after all, had been intended primarily as a means of fixing an explosive charge without detection. On April 17, 1777, an obliging Connecticut governor and council met with the inventor and approved of his new, improved torpedo and promised "to afford him assistance of men, boats, powder, lead & c., as he might want, delivered him without stint."[36]

With the blessings of the Connecticut government, Bushnell waited for the first opportunity to try out his explosive device. As his target he picked HMS *Cerberus,* then stationed in Black Point Bay, between Saybrook and New London. On the night of August 13, 1777, Bushnell set out in a whaleboat with a few soldiers. Two of his mines trailed behind, linked together by a long rope. They were designed to explode on contact. Bushnell stopped upstream of the British vessel and set the mines adrift. He wanted the powder-filled casks to float past the *Cerberus* on opposite sides, with the line getting caught on the bow. In this way at least one of the mines would detonate against the side of the ship. But, as before, the inventor was cursed with bad luck. A crewman on a small schooner lying astern of the *Cerberus* noticed the line and hauled on it. While struggling to bring one of the ponderous mines on board, he triggered the firing mechanism, destroying himself, the schooner, and two of the other three crewmen. The other mine drifted away harmlessly. The frigate escaped unscathed, but British seamen were ordered to be cautious, "as the ingenuity" of the rebels "is singular in their secret modes of mischief."[37]

Failing in Connecticut, Bushnell packed up and went to New Jersey in the fall of 1777. There, with the help of local craftsmen in Bordentown, he further refined his mines. He set them afloat on the Delaware River above Philadelphia, hoping the current would carry them downriver against the British fleet. As luck would have it, the plan misfired. Bushnell's guide supplied misleading information, and the mines were cast off too far upstream. The ropes linking them severed, some mines were destroyed, and the rest bobbed harmlessly downriver. To make matters worse, Bushnell had based his scheme on the assumption that the British vessels would be anchored in the main channel of the Delaware; they were not. One keg did make contact, blowing up a boat with several boys in it that drew too close to investigate, but the British warships were not damaged in the least. Aside from the momentary panic it caused among some British sailors, subsequently ridiculed in Francis Hopkinson's *The Battle of the Kegs*, the incident was soon forgotten.[38]

Bushnell retired to his home in Saybrook, having exhausted his monies and himself. While he sulked in dejection, Connecticut made its one and only effort to inform Congress of his work. In April 1778 Governor Trumbull, long a Bushnell supporter, wrote the Connecticut delegates in Philadelphia and directed them to present Congress with the details of Bushnell's inventions. The delegates wrote back, requesting that an account of the inventor's expenses be forwarded since "his genious ought to be encouraged and Rewarded" by the nation.[39] This brought an end to the move. There is no record of the delegates' having petitioned Congress or of Connecticut having forwarded Bushnell's list of expenses to them.

It is by no means certain that Congress would have voted to aid Bushnell if the Connecticut delegates had submitted a formal petition. It does seem certain, however, that Bushnell would have continued his experiments had he the financial resources. He is partly to blame for the fact that he came no closer to getting congressional aid than he did. There is nothing to indicate he ever took the initiative to petition Congress on his own. Other inventors, notably John Macpherson, were far less bashful. At some point Bushnell should have realized that his ideas would not sell themselves.

Yet the inventor had good reason to be disappointed. Even his stoutest champions did not do all that much to help him. Governor Trumbull, though helpful in securing some funds, might have done more. Instead of relying on others, the governor could have notified Congress about Bushnell himself—such information could have been included in his routine correspondence. The Connecticut Assembly, to its credit, twice granted stipends enabling Bushnell to continue his work. But it too might have done more, considering how much it invested in munitions production, sometimes with no return. All who alluded to Bushnell in their reports and private correspondence did so in flattering terms, blaming his failure on misfortune rather than on conceptual inadequacies. Yet there was never a concerted movement to underwrite Bushnell's experiments. Bushnell's contemporaries treated his devices, and by implication Bushnell himself, as curiosities, not as potentially revolutionary tools of war. As it was, Bushnell's wartime experiments ceased on the Delaware. After returning home he accepted a commission in the sappers and miners, a group he commanded by the war's end.[40]

That Bushnell even got a hearing from patriot leaders and had their support for a time is testimony to the inventive curiosity—and wartime desperation—of his countrymen. But they were in no position to bankroll inventors like Bushnell. They were not accustomed to becoming that involved in invention. It was one thing to promote inventions with a demonstrated usefulness; thus Washington had routinely experimented with new farming equipment at Mount Vernon. It was quite another to promote inventions that had not come as a response to a simple need. Although Washington and the rest may have been intrigued by Bushnell, they made no connection between Bushnell and the larger process of invention, where new devices could revolutionize old ways of doing things.

Americans at this stage in their social development simply did not attach all that much significance to invention, a point that is too often overlooked. Consequently, Roger Burlingame pointed to Bushnell as the epitome of the ingenious Yankee, calling his experiment off Manhattan in August 1776 the typical "Yankee's calm acceptance of the impossible."[41] Bushnell was indeed remarkable, but his individual genius should not be extended, in the manner of Burlingame, to provide proof of a pervasive "Yankee ingenuity" or a nationwide fascination with all mechanical devices. Bushnell proposed projects that pushed the new nation beyond its modest technological limits. Patriots during the war did not exhaust themselves trying to find ways of exploiting Bushnell's genius. They did not follow his every move, because they did not think his experiments were especially useful, and he ended the war an obscure figure.

So did others. One was an acerbic inventor named Joseph Belton, who also designed a self-propelled, submersible "commerce destroyer." Belton in fact sought congressional support for his project before anything was known about Bushnell. Franklin's letter to Silas Deane first mentioning the "Turtle" also alluded to Belton, then in Philadelphia soliciting official patronage, both from the Continental Congress and the Pennsylvania committee of safety.[42]

Hailing originally from Connecticut and a 1769 graduate of the College of Rhode Island, Belton had made Philadelphia his home by August 1775, possibly because he hoped to gain the ear of Congress. Evidently no sketch of his projected submarine survives, but from his own description the device seems to have borne almost no resemblance to the "Turtle." It was far less sophisticated and may have been impossible to build according to the inventor's specifications. Belton's proposed "boat" would have carried a single cannon and moved along in the water with nothing appearing "above the surface much larger than a man's hat." Upon coming within two hundred yards of the target, the entire machine would have submerged, slipped along the ship's keel, and discharged its gun, theoretically blasting a huge hole below the water line. Instead of using Bushnell's system of cranks and paddles, Belton proposed to control his machine by a grappling hook doubling as an anchor and by a finlike apparatus worked by foot. The machine would have been launched upstream of the intended victim; after that the pilot was on his own. He could ascertain his progress by peering out a small pane of glass set in a hinged door.[43]

Belton was confident that a fleet of his submersibles would wreak havoc among British warships in American coastal waters. His listeners were unimpressed, and he failed to generate any official interest. Neither Congress nor Pennsylvania offered help, so Belton did not build his submarine. Apparently he stayed on in Philadelphia anyway, hoping for an opportunity to change a few minds. He managed to obtain a letter of introduction to George Washington from Benjamin Franklin in July 1776, when Franklin informed the general that Belton had petitioned Congress unsuccessfully for assistance. Making an indirect comment on the receptivity of Congress to invention, Franklin wrote that "as they appear to have no great opinion of such proposals, it is not easy, in the multitude of business before them, to get them to bestow any part of their attention on his request." Now desperate to prove himself, Belton only wanted Washington's "countenance and permission" to use his submarine against the British on the Hudson. He would pay his own expenses. Washington had nothing to lose in giving his consent, yet he said nothing more about the proposal, probably because Bushnell was on hand with a far superior and proven design.[44]

Though Belton had been frustrated in one pursuit, Congress had not heard the last from him. On April 11, 1777, he notified Congress that he had invented a process for transforming common single-shot muskets into rapid-fire pieces. He claimed his new muskets could be fired as usual or loaded with eight bullets, these bullets to be fired one after another in a matter of seconds. The innovation had been kept a secret until that time, he explained, so the army could be rearmed in a few months— "which our enemy should know nothing of, till they should be maid to know of it in the field, to their immortal sorrow."[45] Since Congress sought a solution to the perennial munitions problem, this time Belton attracted attention. The Board of War, to whom his petition was referred, returned a favorable report, and Congress authorized Belton to superintend the making or altering of one hundred muskets.[46]

Having been successful to this point, Belton went a step too far for a Congress leery of financial adventure. Seeking to profit from his discovery, he suggested that if three field officers found his claims invalid he should receive nothing; but if they proved valid, he should be awarded a prize of $1,000 by each of the states. He then refused to proceed further until his conditions were met. Shortly thereafter he modified his terms, proposing that if one hundred men armed with his guns proved superior to twice their number armed in the conventional manner, he should get $500 from each state; if three times better, $1,000; and if four times,

$1,500 per state. Congress, not pleased with either proposal, summarily dismissed the petition.[47]

Belton, having gotten his foot inside the door, was not about to give up. He petitioned Congress a second time on July 10, appending an affidavit attesting to the effectiveness of his muskets. The affidavit bore the weighty signatures of Generals Horatio Gates and Benedict Arnold and inventors Charles Willson Peale and David Rittenhouse. Belton reminded Congress that Great Britain made it a policy to support inventors, a practice, he contended, that should be imitated. He also made an insightful comment on the plight of inventors lacking such support. "As money is stild the sinews of war, so it may be stild the sinews of invention," Belton reproved, "for doubtless many experiments which might have discovered something usefull to the Publick, have for want of it died in Obliven."[48]

Despite Belton's plea and his affidavit, Congress refused to budge. The Board of War rejected his petition, and its decision was upheld.[49] Belton, perhaps disgusted with this turn of events, disappeared from sight. He sailed to Europe after the war, where he was equally unsuccessful in securing backing in official circles. He did manage to sell a few of his rapid-fire muskets, by then converted into breechloaders, to the Dutch East India Company.[50]

Congress may have passed up a potentially valuable weapon, but Belton had expected too much. Congress could not impose on the states in the manner he suggested. Moreover, Congress and the states were hard pressed to arm troops with conventional weapons. The retooling, retraining, and rethinking of strategy and tactics required to make maximum use of Belton's invention, even if it was a wonder-weapon, which is doubtful, was too much for a nation hanging on by its fingertips.

Two years passed after the applications of Belton before another inventor approached Congress. And this particular inventor, one of the last to apply to Congress before Yorktown, made contact by proxy rather than in person. Stephen Sayre, merchant, diplomat, and international gadfly, had little choice in his course of action because he spent the war years in Europe. He could not seat himself on the doorstep the way Macpherson and Belton had done.

Born in 1736, Sayre spent his youth on Long Island, his family's home for several generations. He took both his bachelor's and master's degrees at the College of New Jersey and later received an honorary degree from Harvard. Sayre journeyed to England, deciding to make his mark in the world of finance. His good looks and gregariousness endeared him to many, including Charles Townshend and William Pitt. He rose quickly to partnership in the merchant firm of Dennys De Berdt. He later started his own banking house and, in 1773, along with William Lee of Virginia, was elected a sheriff of London, a rare honor for a colonial. He identified himself with the patriot cause quite early and was imprisoned for a while in the Tower of London on a trumped up charge of plotting to seize the King and overthrow the government. With the outbreak of war, Sayre wound up in Paris, where he attached himself to American diplomat Arthur Lee, a London acquaintance. He worked for Lee in Berlin, and when Lee returned to Paris, Sayre leapfrogged from Amsterdam to Copenhagen to Stockholm, trying to drum up support for American commerce and advance his personal fortune. Sayre later took credit for floating the League of Armed Neutrality during his travels, a preposterous boast. Sayre in fact proved to be almost as obnoxious as he was talented. He badgered Benjamin Franklin and others for lucrative offices both during and after the war.[51]

By late 1778 or early 1779 Sayre had devised his "improved gunnery system" and new ship design, which he described to Franklin and a few others. He engaged builders to construct a scaled-down model of his frigate and came close to launching a full-scale version. He first apprised Congress of his innovation in July 1779, but only indirectly. Governor William Livingston forwarded to Congress a letter written to him by Baron Van der Capellen of Amsterdam, in which reference was made to Sayre. Van der Capellen had met with Sayre and was as smitten by him as John Adams was by John Macpherson. He reported that Sayre had perfected a way to make stronger, lighter, swifter, and cheaper merchantmen. Even more important, Sayre had developed an original method of building warships by "a plan I believe never before entered into the Imagination of Man." The Dutchman repeated Sayre's claim "that the power which shall first introduce it into the marine will be able to annihilate that of its Enemies." He said no more, closing with the note that Sayre intended to appear before Congress and present his case in person.[52]

Congress took no notice of the letter and simply filed it away. Perhaps memories of John Macpherson and Joseph Belton were too fresh in some minds. Livingston went ahead and mentioned the Van der Capellen letter and news of Sayre to Washington. Although Washington expressed an interest in the "extraordinary" plan, there was little he could do with its author overseas.[53] Sayre, for his part, did not return to his native country until after the war. Thus it was that serious deliberation on Sayre's proposal did not take place until long after the fighting had ceased. Nevertheless, Sayre's combined fast frigate-improved gunnery system is worth investigating in connection with the war itself. On paper it sounded immensely innovative, the only invention to rival Bushnell's "Turtle" in significance. Sayre detailed his plan in an anonymous pamphlet, *Cursory Observations Relative to the Mounting of Cannon in a New Way*.[54] The pamphlet was not published until 1785, when Congress was considering Sayre's request for compensation for his diplomatic service.

Chances are that Sayre had had the same basic plan in mind six years earlier.

Apparently Sayre had by 1785 constructed a rough model of his newly designed gun carriage for public perusal. Those who saw the model, however, were not favorably impressed, so Sayre wrote his pamphlet to sway a skeptical public as well as Congress. His gun carriage embodied a new approach to artillery design and function. He called for an early form of continuous aimed firing, this system to replace the common fixed gun and carriage fired by a match held at arm's length. Pinpoint accuracy was next to impossible with common naval cannon, which had a restricted field of fire and could not be adjusted to compensate for a ship's natural dip and roll. Sayre proposed modifications, with the gunner being seated on the gun carriage itself, which would be trimmed down from the usual dimensions. The inventor was convinced gun recoil would not harm the gunner, who would sight along the barrel. Though the validity of this part of Sayre's theory is questionable, he is on safer ground hereafter. He suggested that the trunnions holding the cannon on the carriage be redesigned to facilitate moving the gun barrel up and down. Sayre also suggested that his new gun carriage be mounted on a swivel for side-to-side movement and perhaps be placed in tandem or in fours for greater concentration of fire.

According to Sayre's blueprint, ships would need fewer guns and could be smaller and thus faster, more maneuverable, and harder to hit. Sayre calculated that a ship adopting his design and armed with twenty-eight guns would be the match of a seventy-four-gun ship of the line. He emphasized how the superior concentration of fire possible under his system made existing coast defenses and warships obsolete. Naturally he implied that Congress had greatly erred by not taking an active interest in his plan during the war.

Sayre petitioned Congress under an alias in 1785, apparently fearing that he would lessen his chances of success if Congress joined the naval gunnery plan with his controversial request for payment for wartime diplomatic service.[55] The committee to whom Sayre's gunnery proposal was referred recommended that secretary at war Henry Knox judge the merits of the project. If Knox approved the plan, then the inventor would be granted "a reward adequate to its importance and utility."[56] After meeting with Sayre, Knox reported that the inventor wanted a substantial sum. Congress balked at this and filed Sayre's petition, the committee recommendation, and Knox's findings without further comment.[57]

Sayre suffered the same fate as Belton. His desire to receive remuneration commensurate with the presumed importance of his discovery was too much for a budget-conscious Congress. Like Belton, Sayre did not give up. He continued to promote his scheme, concentrating his efforts on Knox and George Washington instead of Congress.[58] Wise enough to see that such patronage, if obtained, would prove indispensable, Sayre eventually became irritated with Knox when he did not succeed.[59] Knox had initially been encouraging, but he advised Sayre in 1785 to wait until the federal government was strengthened. Five years later, when Knox failed to prosecute his case, Sayre believed that he had been misled at best, betrayed at worst. After numerous inquiries from the disgruntled Sayre, Knox finally responded that "every effort of Genius, capable of demonstration and which will either directly or indirectly promote the power and interest of the United States is justly entitled to reward."[60] So until Sayre gave a practical demonstration of his invention, Congress could not be persuaded to lend financial support for research and development, regardless of how good it sounded in theory. Sayre remained unwilling or perhaps technically unable to execute his designs without money. His new cannon was never built, and the plan itself was pigeonholed and soon faded from memory.

Sayre, Belton, Bushnell, and Macpherson were not the only inventors to come forward with unusual plans during the war. John Stevens of New Jersey, later to gain fame for his steam engine and railroad work, also dabbled in invention during the Revolution.[61] In 1778 he secured from Washington permission to try out a "machine" he had designed for destroying British ships on the Hudson. He never made this device, however. Washington at first agreed to provide Stevens with men and materials but later changed his mind after learning that workmen could not be spared and that some members of his staff scoffed at Stevens's plan. Foreshadowing the stance Congress took with Sayre after the war, Washington concluded that "the probability of its answering the end proposed shall be well considered before it is carried into execution."[62]

Captain Ephraim Anderson of the New Jersey line approached Congress in July 1776 with plans for specially constructed fireships to be employed against the British fleet around New York.[63] Anderson's overall plan was hardly unique. Fireships, like row galleys, were commonly used in the war. But there was a trick to building these floating tinderboxes so that they would burn quickly and evenly, allowing the pilot to stay on board until the last second. Anderson had sought to interest New York in his particular type of fireship a year earlier but to no avail. He also contacted the Pennsylvania committee of safety, where he had the satisfaction of seeing some of his ideas put into practice. With Congress he was even more fortunate, receiving permission to pursue his plan. John Hancock, then president of Congress, wrote Washington that "Events only can shew whether his Scheme is visionary or practicable." The very chance of success made the undertaking worthwhile, Hancock continued, and should "it fail, our Situation will be, in every Respect the same as before."[64] Washington, although skeptical, cooperated and detached fifty artificers to assist Anderson.[65] Anderson kept Congress posted on his progress, prophesying that he would demolish the enemy fleet.[66] His fireships did enjoy some success against a few small ships, but they did nothing to alter or even impede the campaign which netted Manhattan Island for the British. Anderson's career ended in 1777, when he died of natural causes.

Sir James Jay, older brother of John Jay, also made a foray into wartime invention. A practicing physician, Jay had commercial investments abroad. While in England before the war, he had studied the techniques of casting and boring cannon. He wanted to test the range of cannon at West Point, for what reason it is unclear. Washington granted him permission in November 1778 to run his tests. Jay planned to conduct his "experiments" in the dead of winter when the Hudson was frozen over and he could accurately mark where the shot struck after being fired. Major Sebastian Bauman had permission to conduct similar experiments, independent of Jay, though it does not appear that either man went through with his respective proposal.[67]

Neither Jay nor John Stevens made any sustained effort to garner congressional backing. Anderson died too early in the war to tell whether he possessed inventive expertise in anything other than the building of fireships. Still, including them in the total, only a very few inventors applied to the Continental Congress or approached George Washington over the course of the war.

When I. Bernard Cohen wrote in 1945 that the Revolution "marked the first union of the forces of science and technology with government," he may have been influenced too much by the close interaction of science and government in World War II.[68] Cohen echoed Thomas Paine who had claimed that the war "energized invention and lessened the catalogue of impossibilities."[69] Paine, like Cohen later, overstated his case. Brooke Hindle was much closer to the truth in contending that the "disruptive influence of the war upon the whole pattern of science in America was much more serious than the limited number of beneficial influences it provided."[70] The war had not produced a magical melding of science and technology. American inventive ingenuity had less to do with bringing victory than British weariness and the French alliance.

Except for the April 1777 resolution asking the states to keep it posted on inventive activity, Congress did not take the initiative in stimulating invention. It was not set up to promote inventive inquiry and had no funds for the eighteenth-century version of research and development. Congress only reflected the larger tendencies of a society unused to promoting invention. In part this may have been an outgrowth of colonial-imperial relations; in part it may have resulted from the as yet unclear connection between invention, technological innovation, and social change. It is significant, if not surprising, that almost none of the inventions brought to the attention of Congress were concerned with upgrading modes of industrial production.[71]

Of the handful of unusual inventions made public, only those of Bushnell, Belton, and Sayre were particularly innovative or capable of exerting an impact on the war. Even then, neither the "Turtle" nor Sayre's gunnery system was so revolutionary as to change the nature of the conflict, even if they might have forced a change in naval tactics. Sayre's gunnery system, like Belton's musket, posed too heavy an expense for the slim resources of Congress and the states. It is possible that Sayre could not have delivered on his promises anyway since, unlike Belton and Bushnell, he never proved his theory in practice. Only David Bushnell's ideas were practical enough to have enjoyed a true chance of success. Nevertheless, with Bushnell, as with other inventors, can be seen the American perception of invention as a source of curious gadgets or possibly useful tools, but not as the source of important change.[72]

Inventors like Bushnell, who hoped to introduce dramatically new devices, failed. It would be well over a century before the national government and inventors began to work smoothly in concert.[73] Until then, the more unique the device, the less likely it was to be used. Nonetheless, inventors during the War of Independence were not always unsuccessful. Those who had more modest ambitions, who designed new devices that could be easily adapted to a previously defined need, often succeeded. Ingenuity had not been swept away during the war; rather, ingenuity had to be channeled into certain types of projects to prosper. Those projects helped keep Yankee ingenuity alive.

NOTES

1. Hindle, *Pursuit of Science*, pp. 220-233.

2. Hobday in McIlwaine, ed., *Journals of the Council*, I, 77, II, 6; Colles in Ristow, ed., *Survey of the Roads*, pp. 16-20; and Force, ed., *Amer. Archives*, 4th series, III, 259, 284, for Colles's unsuccessful attempt to erect a wool card wireworks.

3. Boyd, ed., *Papers of Jefferson*, II, 202-203.

4. *Va. Gazette* (Purdie), 16 May 1777.

5. Charles F. Adams, ed., *The Works of John Adams*, 10 vols. (Boston: Little, Brown and Co., 1850-1856), II, 448. For a more judicious assessment, see H. James Henderson, *Party Politics in the Continental Congress* (New York: McGraw Hill Book Co., 1974).

6. Ford, ed., *JCC*, VII, 291.

7. Jonathan T. Lincoln, "The Beginnings of the Machine Age in New England: David Wilkinson of Pawtucket," *New England Quarterly* 6 (1933):720; Bishop, *American Manufactures*, I, 504, refers to a Jeremiah Wilkinson of Cumberland, Rhode Island, who did the same thing back around 1775.

8. Foner, ed., *Writings of Thomas Paine*, II, 1135-1136. Sion Seabury's design for a "rolling breastwork" would have been impractical—see Seabury's letter to Ezra Stiles in Dexter, ed., *Literary Diary of Ezra Stiles*, I, 497-498.

9. See Macpherson's unfinished autobiography, *A History of the Life, Very Strange Adventures, and Works of Captain John Macpherson* (Philadelphia: Zachariah Poulson, 1789); and William Macpherson Hornor, "Extracts from the Letters of John Macpherson, Jr., to William Patterson, 1766-1773," *PMHB*, 23 (1899):51-52.

10. John Macpherson, *Letters* (Philadelphia: William Evitt, 1770); idem, *Letter to John Dickinson, Esq.* (Philadelphia: Robert Bell, 1770); and idem, *A Pennsylvania Sailor's Letters, Alias the Farmer's Fall* (Philadelphia: Robert Bell, 1771).

11. Hazard, ed., *Pa. Archives*, 8th series, VIII, 6605.

12. Adams, ed., *Works of John Adams*, II, 424, 428.

13. Ford, ed., *JCC*, III, 296, 300, 301; Hancock letter to Washington is in the GWP, series 4, letter of 20 October 1775 (reel 34).

14. "Warren-Adams Letters," *Mass. Hist. Soc. Coll.*, LXXII, 156, 168-169, 177.

15. Fitzpatrick, ed., *Writings of Washington*, IV, 71-72, 76; William Reed, ed., *Life and Correspondence of Joseph Reed*, 2 vols. (Philadelphia: Lindsay and Blakiston, 1847), I, 126; and the GWP, series 3A, I, 66-67 (reel 14), letter from Washington to John Hancock of 8 November 1775.

16. Josiah Quincy in the Adams Papers (Boston: Massachusetts Historical Society, 1958, 608 reels microfilm), I, 86 (reel 345), letter to John Adams of 22 September 1775; also see the GWP, series 4 (reel 34), letter to Washington of 31 October 1775; for Paine see Foner, ed., *Writings of Thomas Paine*, II, 1067-1077.

17. PCC, Item 78, XV, 23-24, 29-30, 53, 54 (reel 99).

18. Ibid., Item 78, XV, 321 (reel 99); Item 42, V, 144, 150, 154 (reel 55); Item 147, III, 541 (reel 158); Ford, ed., *JCC*, XIV, 859. Recommendations to Pennsylvania in Hazard, ed., *Pa. Archives*, 1st series, VII, 180. The request for a naval commission is in the PCC, Item 41, VI, 207, 210 (reel 51).

19. Frederick Wagner, *Submarine Fighter of the American Revolution* (New York: Dodd, Mead and Company, 1963), p. ix.

20. Biographical details in ibid., pp. 5-22; and David Humphreys, *An Essay on the Life of the Honorable Major General Israel Putnam* (Boston: Samuel Avery, 1818), 108n. The eulogy is in Timothy Dwight, *Greenfield Hill* (New York: Childs and Swaine, 1794), Book VI, pp. 161-162, lines 423-436.

21. See David Thompson, "David Bushnell and the First American Submarine," *United States Naval Institute Proceedings* 68 (1942):176-179; and Alex Roland, "Bushnell's Submarine: American Original or European Import?" *Technology and Culture* 18 (1977):157-174, for comments on forerunners of Bushnell whose writings may have influenced him. Also see idem, *Underwater Warfare in the Age of Sail* (Indianapolis: Indiana University Press, 1978), pp. 62-88; and Robert Speck, "The Connecticut Water Machine Versus the Royal Navy," *American Heritage* 32 (December 1980):33-38.

22. Described in the *APS Trans.* 4 (1799):303-308; also in Charles Griswold, "Description of a Machine," *American Journal of Science and Arts* 2 (1820):94-98.

23. Benjamin Gale to Benjamin Franklin, 7 August 1775, in the Franklin Papers, IV, fo. 61, APS MSS.

24. Dexter, ed., *Ezra Stiles*, I, 600.

25. Letter to Deane from Gale of 10 August 1775, in the *Connecticut Historical Society Collections* II, 294.

26. Letter to Deane of 9 November 1775, in ibid., II, 315-317.

27. Gale to Franklin, 7 August 1775, Franklin Papers, IV, fo. 61, APS MSS.; Franklin to Deane in Labaree, ed., *Papers of Franklin*, XXII, 183-185, letter of 27 August 1775.

28. Adams Papers, I, 103 (reel 345), letter of 23 October 1775.

29. Benjamin Gale to Silas Deane, 22 November 1775, in the *Conn. Hist. Soc. Coll.*, II, 322, 333-334.

30. Hoadly, ed., *Conn. Public Records*, XV, 233-236.

31. Burnett, ed., *Letters*, II, 42.

32. Charles S. Hall, *Life and Letters of Samuel Holden Parsons* (New York: Otseningo Press, 1905), p. 60.

33. Quote from James Thacher, *Military Journal* (Boston: Richardson and Lord, 1823), p. 76; Ezra Lee's account is in Henry P. Johnston, ed., "Sergeant Lee's Experience with Bushnell's Submarine Torpedo," *Magazine of History* 29 (1893):263-266; Bushnell's account is in the *APS Trans.* 4 (1799):310-311.

34. Fitzpatrick, ed., *Writings of Washington*, XXVIII, 278-281.

35. The quote is from enclosures sent from Bushnell to Ezra Smith in October 1787, in *Naval Documents*, VI, 1506-1507, appendix B; also see William Heath, *Memoirs* (Boston: J. Thomas and E. Andrews, 1798), p. 61.

36. Hinman, ed., *Historical Collection*, p. 437.

37. An excellent collection of Bushnell documents is in Henry Abbot, ed., *The Beginnings of Modern Submarine Warfare* (Willets Point: Engineer School of Application, 1881); see pp. 40-41 for the statement by the *Cerberus* commander. Also see the accounts in Thacher, *Military Journal*, pp. 149-150; and Bushnell's personal recollections in the *APS Trans.* 4 (1799):311-312.

38. Wagner, *Submarine Fighter*, pp. 84-86.

39. Hinman, ed., *Historical Collection*, p. 531; Burnett, ed., *Letters*, III, 202.

40. The sappers and miners commission recommendation is in a letter from Jonathan Trumbull to George Washington of 29 May 1779 in the GWP, series 4 (reel 59).

41. Burlingame, *March of the Iron Men*, p. 146.

42. Labaree, ed., *Papers of Franklin*, XXII, 185.

43. Hazard, ed., *Pa. Archives*, 1st series, IV, 650-652. Belton is mentioned in Labaree, ed., *Papers of Franklin*, XXII, 185n.

44. Burnett, ed., *Letters*, II, 20; GWP, series 4 (reel 35), letter of 22 July 1776.

45. PCC, Item 41, I, 123 (reel 48).

46. Ford, ed., *JCC*, VII, 324.

47. PCC, Item 78, II, 75-77, 183 (reel 91); Ford, ed., *JCC*, VII, 361.

48. Ibid., Item 42, I, 138-139 (reel 53).

49. Ibid., Item 147, I, 278 (reel 157); Ford, ed., *JCC*, VIII, 542, 566.

50. Howard Blackmore, *British Military Firearms, 1650-1850* (London: Herbert Jenkins, 1961), p. 249.

51. Stephen Sayre, *A Short Narrative of the Life and Character of Stephen Sayre* (no. pub., 1794); John Richard Alden, *Stephen Sayre* (Baton Rouge: Louisiana State University Press, 1983), pp. 105-121, 154-156, tells most of the ship design

story. See the Franklin Papers, APS, for Sayre's correspondence with Franklin, particularly the letters from Sayre to Franklin of 7 November 1778 and 21 March 1779. Interestingly enough, Franklin and Deane noted in a letter to Congress of 26 May 1777 (PCC, Item 85, 57-60, reel 114) that a frigate was being built in Amsterdam reputedly the equal of a seventy-four-gun ship. They also recommended that fast frigates raid the English and Scottish coasts. This all sounded very much like Sayre's plan of 1778 and 1779 (although Deane had urged that shipping raids be made along the English coast as early as 1776). Perhaps Franklin and Deane had talked with Sayre about ship design and coastal raids as early as the spring of 1777, before Sayre went to Copenhagen and studied ship design innovations being made there.

52. PCC, Item 68, 494 (reel 82).

53. Fitzpatrick, ed., *Writings of Washington*, XVII, 225.

54. New York: Eleazer Oswald, 1785.

55. PCC, Item 78, I, 459 (reel 90); Ford, ed., *JCC*, XXVIII, 346n.

56. Ibid., Item 19, V, 287 (reel 28); Ford, ed., *JCC*, XXVIII, 395.

57. Knox report on 19 May 1785 in the PCC, Item 151, 17 (reel 165); report filed note in Ford, ed., *JCC*, XXVIII, 395n.

58. Sayre's letter to Henry Knox is in the Knox Papers, XXII, 152 (reel 22), 3 October 1788; also see the letters of 3 January 1789, XXIII, 52 (reel 23); and to Washington of the same date in the GWP, series 7, 25-28 (reel 123).

59. Knox Papers, XXVI, 76 (reel 26), letter of 15 June 1790.

60. Ibid., XXVI, 174 (reel 26), letter of 4 September 1790; final letter from Sayre in ibid., XXXIV, 177 (reel 34). Knox, who was also a patent commissioner, may have thought that active support of Sayre was unethical.

61. For Stevens see Archibald Turnbull, *John Stevens: An American Record* (New York: The Century Co., 1928); and Carl Mitman, "John Stevens," in Dumas Malone, ed., *Dictionary of American Biography*, 20 vols. (New York: Charles Scribners's Sons, 1928-1937), XVII, 614-616; neither mentions Stevens's wartime inventive activity.

62. Fitzpatrick, ed., *Writings of Washington*, XII, 314-315; Washington's statement is in the GWP, series 4 (reel 51), letter to William Malcolm of 24 August 1778.

63. Anderson to Washington on 9 July 1776 in the GWP, series 4 (reel 35). Anderson had failed to interest New York in his fireships; see his letter to the New York committee of safety of 4 November 1775 in the *Journals of the Provincial Congress*, I, 200. Anderson was slightly more successful in Pennsylvania; see the next chapter. He was also apparently the pilot of a fireship used at Quebec in May 1776. See Thomas Ainslee, "Journal," in Sheldon S. Cohen, ed., *Canada Preserved* (New York: New York University Press, 1968), pp. 87-88. Anderson's wartime service is summarized in William Stryker, ed., *Official Register of the Officers and Men of New Jersey in the Revolutionary War* (Trenton: William T. Nicholson and Co., 1872). pp. 16, 32, 77.

64. Burnett, ed., *Letters*, II, 8.

65. Fitzpatrick, ed., *Writings of Washington*, V, 275, 286-287, 300, 343-344.

66. PCC, Item 78, I, 11, 19 (reel 90).

67. Fitzpatrick, ed., *Writings of Washington*, XIII, 360; XVII, 442, 454; XVIII, 332, 350. Jay to Washington in the GWP, series 4 (reel 63) letter of 21 January

1780. A note on his observations of British cannon making is in the PCC, Item 78, XIII, 171-174 (reel 98), letter to Congress of 18 July 1780.

68. Cohen, "Science and the Revolution," pp. 367-368, 374, 376, 378.

69. Foner, ed., *Writings of Paine*, II, 1135-1136.

70. Hindle, *Pursuit of Science*, p. 247; similar statement in Bell, "Scientific Environment of Philadelphia," pp. 12-13.

71. There was at least one exception to this generalization. In March 1777 Congress contracted with a Samuel Wheeler "for a number of cannon by the new construction"; see Ford, ed., *JCC*, VII, 193, 228, 272.

72. Major Bauman should perhaps not be included in this list, since his experiments were evidently not part of a plan to make some new type of cannon. James Hopkins could be added to the list. He approached Congress late in 1781, as the war was winding down—see chapter 8. Samuel Wheeler, mentioned earlier, did not claim to have invented the mode of cannon construction himself. Dresden merchant Samuel Golden sought congressional patronage for his proposed road improvements and mining methods, which he wanted to introduce in America, but he was unsuccessful (PCC, Item 78, X, 103-107, reel 95).

73. William McNeill, *The Pursuit of Power* (Chicago: University of Chicago Press, 1982), p. 224. Also see Paul C. Koistinen, "The 'Industrial-Military Complex' in Historial Perspective: World War I," *Business History Review* 41 (1967):378-403.

5 GENIUS RECOGNIZED: THE ENGINEERS

From April 1775 on, the patriots scurried feverishly to meet the demands of war. Engineering played an essential role in their jumbled war plans. Since the conflict became a war of containment as well as movement, of stationary defense and mobile offense, engineers were nearly as indispensable as line troops. To the engineers fell the job of designing forts, erecting breastworks, and laying out camps. They also worked on a few somewhat more exotic projects. Those engineering projects, from the great chains that spanned the Hudson River to the multiple defenses along the Delaware River, opened opportunities for the ingenious. The response was enthusiastic, the results, impressive. Ingenious craftsmen rendered valuable service, service that did not go unnoticed and was not soon forgotten.

One of the more urgent engineering projects entailed erecting defenses for major cities along the Atlantic seaboard. These ports, and the rivers and bays opening them to maritime trade, were strategically crucial. Their importance in keeping alive commercial and diplomatic relations with Europe and the West Indies made them likely targets for a British attack. Consequently, Boston, New York, Philadelphia, and Charleston initiated building programs to secure themselves against naval assault. Local residents, sometimes with, and more often without, the aid of professional engineers, threw up defensive works in a flurry of activity. As an added precaution, they sank obstructions in the Hudson and Delaware rivers and in the harbors of Boston and Charleston.

The success of these measures depended to some degree on technical knowledge, knowledge that is the stock in trade of a professional military engineer. While military engineers were common in Europe, the colonies had barely a handful in 1775. A chronic shortage of trained engineers in fact plagued the patriots throughout the War of Independence. The complications arising from this shortage were most evident during the first two years of the war, before the arrival of professional European engineers. Until that time, the Americans relied on their own makeshift techniques.

George Washington's complaints in the summer of 1775 about the dearth of engineers in his army at Cambridge are by now familiar. Those few on hand lacked formal training and knew practically nothing about field works and fortifications. Washington's request that Congress recruit qualified engineers for service in the Continental army was a difficult order to fill. Benjamin Harrison advised him not to be too optimistic, for "the want of Engineers I fear is not to be supplied in America."[1] Still, the outlook was not as bleak as could be expected. Though professional engineers were scarce, colonial urban centers possessed an abundance of skilled craftsmen. These shipwrights, mechanics, ironworkers, and carpenters formed the backbone of the construction trades. They constituted a respected class in the urban community, people whose talents were indispensable. In war as in peace they were a vital reservoir of practical expertise. Colonial assemblies harnessed their energy and channeled it into various projects, notably the building of defensive works.

Pennsylvania in particular made good use of its innovative artisans. Carpenters, smiths, and shipwrights pooled their resources and bent their collective shoulders to the task of fortifying the Delaware River. On the surface it appears that all their efforts were wasted, for Philadelphia fell to the British in 1777. Upon deeper investigation, however, it could be argued that those craftsmen were remarkably successful. They skillfully executed many of their designs, relying on expertise developed in their respective trades. Their successes showed that Yankee ingenuity could be effectively mobilized for war.

They understandably enjoyed more success in some areas than in others. Their failings illuminated certain blank spots in colonial technical knowledge, echoing a long-standing dependence on the transit of technology. Native genius and trade skills could only be taken so far; after that a more refined skill born of formal schooling and experience was needed. Neither the mother country nor the colonies had a recognized community of civil engineers, but the mother country did have professional military engineers. The colonies, with few exceptions, did not. But then there had been no real need for American-born, American-trained military engineers before the War of Independence. During the wars for empire, the British had customarily brought along their own. Those colonists who worked as engineers during the War of Independence accordingly learned as they went. Since Pennsylvania "engineers" called into service in 1775 lacked the advantage of rigorous training, defense problems taxed them to the fullest.

Nathanael Greene once remarked that "Philadelphia is the American Diana, she must be preserved at all events."[2] Philadelphia's strategic importance notwithstanding, in the spring of 1775 the Delaware River flowed past virtually undefended shores, and the city lay exposed to a potential British naval sortie. Wealthy Philadelphia lawyer Joseph Reed therefore cautioned the Pennsylvania General Assembly in June that defense measures for the Delaware warranted close attention. "Vain are all our Military Operations," he warned, "if a single Ship of War may proceed, without any Interruption, to the very Front of the City."[3] Another citizen wrote, in a bit of understatement, "If any of the King's Ships should come up to this City, there is reason to apprehend it will be with hostile Intentions." Thus, he surmised, "It might not be improper to prevent the attempt."[4]

Aware of this danger, on June 30 the Assembly created a twenty-five-man committee of safety whose responsibilities included providing for defenses in and along the Delaware. One of the committee's first acts was to appeal to local craftsmen, inviting them to offer solutions to the problem of Delaware defense.[5] Pennsylvania, like most of her sister colonies, had no military engineers, so it fell back on those craftsmen answering the committee of safety's call for aid. Although craftsmen enlisted in Pennsylvania's service were not military men by vocation, of all the professions in the colonies their training most closely approximated that of engineers. They adapted their accumulated learning, gleaned from trade skills and practical experience, to designing defenses for the Delaware.[6]

Putting the Delaware into a defensible state required a fairly complicated master plan, because it was not enough merely to toss up earthworks at selected points along the river. A successful defense network would have to be fully integrated, with land batteries laid out to cover obstructions in the river itself. Those obstructions in turn needed to be supported by a naval force. To close the Delaware to enemy traffic, the patriots had to first pick the most defensible stretch of the river between Philadelphia and the Delaware capes. Next, they had to design land and water obstacles and assemble a flotilla which together would deter the British. All of this needed to be done promptly, for there was no predicting when the enemy would turn up at the mouth of the Delaware.

Some craftsmen devised obstacles for the main channel of the river. Lewis Nicola, a Philadelphia merchant of modest means most remembered for his minor part in the so-called "Newburgh Conspiracy" near the end of the war, was one of the first to contact the committee of safety. Nicola had studied engineering, the military arts, and science in general for sometime. An accomplished surveyor, he also owned a circulating library and had edited the short-lived *The American Magazine*. For a man of scientific pretensions, however, his river defense proposal was rather whimsical. Nicola called for the construction of two large warships of eighty guns apiece. These ponderous vessels would be designed so that when turned to straddle the main channel all of their guns faced downriver, presenting a full broadside of overwhelming power. Pig iron stored on the unarmed sides of the ships would balance their weight. Nicola recommended that all the top masts, yards, and cloth be removed once the warships were anchored, leaving only the higher crow's nests, from which vantage points sharpshooters could sweep the decks of the enemy.

The committee of safety thanked Nicola politely for his suggestion but rejected the scheme.[7] It was undoubtedly reluctant to invest in a shipbuilding project consuming considerable time and money in what would be an exercise in futility. Nicola's scheme left too much to chance. The British could, with little trouble, cut the anchor cables and loose the ships from their moorings or destroy them with red-hot shot and move upriver unopposed.

Pennsylvania needed something more substantial than what Nicola had in mind to complement the tiny state navy and the forts being built along the Delaware. Nicola's proposal was too simple, avoiding rather than confronting the engineering difficulties involved in blocking the river. A viable defense required obstructions which could be positioned in the relatively narrow main channel and not be easily disposed of by the British.[8] These obstructions had to be sufficiently solid to withstand wave action and the pull of underwater currents and winter ice and yet be manageable enough for towing into place. In addition, accurate soundings of the river bottom had to be taken at frequent intervals to insure that the obstacles were sunk in the deepest part of the channel and that spaces were not left so large that an enemy warship might slip past. (Twenty years had passed since Joshua Fisher, a Philadelphia merchant, made his hydrographic chart of the river; new soundings were needed.) It was imperative that these barriers be placed effectively because they would stand as the passkey to the overall defense system. Pennsylvania's navy could not hope to win a ship-to-ship duel with the opposition. Therefore the river itself had to be closed off to keep the British from making a dash for Philadelphia.

Several local craftsmen came forward with solutions to the river obstruction conundrum that were more practical than the Nicola scheme. Richard Wells of Burlington, New Jersey, noted inventor and member of the American Philosophical Society, suggested that huge oak piles capped with iron be driven into the river bed, from eight to ten feet apart, to form an underwater wall. Wells designed these piles so that once imbedded they could not be removed. They would stand a few feet below the surface, with broad, flat "trunnels" to prevent their being sunk deeper, and barblike appendages to prevent their being pried up. A passage for friendly ships would be left, which could be closed quickly by sinking a small boat filled with rocks. Ebenezer Robinson, Philadelphia mechanic and inventor in 1774 of a "pump for leaky vessels," advocated something entirely different. His defenses consisted of large, floating timbers joined by chain links. The timbers would be held suspended just below the surface by anchors set diagonally as well as upriver and downriver of the boom. Inventor and committee of safety member Arthur Donaldson, whose "clamshell dredge" had favorably impressed some members of the American Philosophical Society a few months previously, designed a boom slightly different from Robinson's. His consisted of buoys filled with combustibles joined by hawsers anchored in the river at one end and to stone piers at the end nearer shore. Like Robinson and Wells, Donaldson submitted his plan replete with sketches.[9]

All of the foregoing, though not infeasible, were rejected in favor of a superior design presented by Robert Smith, a craftsman of wide reputation.[10] Smith was born in Glasgow, Scotland, in 1722. Very little is known of his early years or at precisely what time he and his family emigrated to Pennsylvania and took up residence in Chester County. Smith became a successful carpenter at an early age and joined the prestigious Carpenters Company, most likely in the late 1740s. By the 1750s he was respected as a master carpenter who has since been labelled "by all odds . . . the most distinguished architect in Philadelphia in the middle eighteenth century."[11]

Smith's first recorded architectural commission was for Nassau Hall at the College of New Jersey in 1754. His plain Georgian style for that building became a popular pattern for buildings on other college campuses, most notably at the College of Rhode Island and Dartmouth. Smith had a hand in designing and erecting a number of famous Philadelphia structures, including St. Peter's Church (1758), Benjamin Franklin's house (1765), Zion Lutheran Church (1766), Carpenters' Hall (1770), and the Walnut Street Prison (1773). In 1769, the same year in which he was made a member of the American Philosophical Society, Smith designed a bridge supported by stone abutments and covered with wood for spanning the Schuylkill River. Smith and other backers of the bridge fruitlessly petitioned the General Assembly for funds until the spring of 1775. Smith never constructed the bridge, but the innovativeness of his design, added to his work on the Delaware, prompted one historian to insist that the Quaker artisan has earned the title "engineer" as well as architect.[12] Smith also became an early convert to the patriot cause and in 1774 signed a circular from the mechanics association condemning the closing of Boston Harbor to commerce.[13]

On July 24, 1775, Smith appeared before the Pennsylvania committee of safety "with a model of a machine for obstructing the Navigation of the River Delaware and explained the Construction of it." The committee accepted the design and Smith's offer to supervise, without pay, the making of an unspecified number of machines.[14]

There has been a minor debate over whether Smith actually originated the machine design himself or borrowed it from someone else. Indeed, various devices for blocking river passages were known and used by military engineers long before the War of Independence, so Smith could make no claim to originality on that score. A few writers insist that Benjamin Franklin, not Smith, deserves credit for coming up with the machines deployed in the Delaware. Carl Van Doren, Franklin's biographer, stated equivocally that the machines "may or may not have been devised by him."[15] Although Franklin championed the Delaware defense project from its inception, it seems improbable that he made the machines used on it. He served as president of the committee of safety at the time of Smith's petition, and it is unlikely that Smith disingenuously claimed the design as his own in Franklin's presence. Instead, it seems more plausible that Franklin, already acquainted with Smith through business transactions and by association in the American Philosophical Society, simply gave Smith the benefit of his general knowledge. Smith may have incorporated some of that knowledge with information culled from other sources, but the end product was his creation.

Smith's contemporaries initially referred to his devices simply as "machines" and later as "bugbears," "caissons," "stackadoes," and most commonly as "chevaux-de-frise."[16] Chevaux-de-frise translated literally is "Friseland horses," after barriers made by the people of Friseland (eastern Holland) to impede cavalry.[17] The Delaware chevaux-de-frise were all of the same basic design, though the carpenters engaged by Smith—such as Joseph Govett, a fellow member of the Carpenters Company—often altered their size or the number or incline of their points, depending on their intended location. For example, the main shipping channel of the river varied in width and depth, with the passage near Billingsport, New Jersey, being narrower and deeper than the channel between Mud Island and the New Jersey shore farther upriver.

Still, the typical chevaux-de-frise took from twenty-five to thirty logs, ranging from forty to sixty-five feet long and from twelve to twenty-six inches thick. Carpenters cut the logs to construct large frames roughly seventeen feet long and almost as wide, and close to fourteen feet high. They lined the inside of the box made by joining these logs, including the floor, with two-inch-thick pine planks. These planks were caulked to make the whole structure as watertight as possible. Two logs over twenty-five feet long and tipped with iron were seated firmly at the bottom edge of both rear corners, from which they jutted out diagonally. When sunk, the tips of those logs ideally lay four to eight feet below the surface at low tide.[18] As a British witness noted later, the completed frames were of such a "prodigious weight and strength" they "could not fail to effect the destruction of any vessel which came upon them."[19]

Another observer commented that they were well balanced by their dimensions, for they could "neither be broken, nor forced backward, nor turned over."[20]

Floating these large and rather unwieldy boxes into proper position was no simple operation. The long list of items necessary for the enterprise included several anchors weighing from 1,100 to 1,400 pounds, towlines, lighter ropes, two boats equipped with windlasses, and a six- or eight-oared galley.[21] Once the chevaux-de-frise were maneuvered to the right spot, Smith and his crew removed special plugs, and the devices were sunk by the inrushing water and anywhere from fifteen to twenty tons of stones lowered inside their bins. Workmen hoisted and dropped the stones into place with a machine Smith made expressly for that purpose. Small anchors tied to the chevaux-de-frise added to their stability.[22]

While Smith busily supervised the building of his machines and waited for word to tow them out into the river, Benjamin Franklin and several other members of the Pennsylvania committee of safety met with the committee of safety from Gloucester County, New Jersey. The Gloucester committeemen had agreed earlier to make Delaware defense a joint endeavor. They helped the Pennsylvanians by assigning a dozen of their people to provide logs and thousands of board feet of pine scantling. Most of the men in both groups were artisans or merchants familiar with the river because of commercial pursuits—such as Robert Whyte, a Philadelphia importer-exporter. A few, notably Franklin and mathematician Owen Biddle, were men of broad scientific interests. Together they chose the location for Smith's obstructions.

The two groups decided to sink three staggered, serpentine rows of obstacles in the river's main channel, about seven miles below Philadelphia and just above the tip of Hog Island, a few hundred yards off Mud Island.[23] They wanted Smith to space the chevaux-de-frise so that the removal of one in the front row would not permit the enemy to pass unhindered through the other two rows. The arrangement of obstructions and choice of locations proved wise, for although the river was roughly two and a half miles wide at this point, the main channel was less than two hundred yards in breadth. And the channel angled close by Mud Island, where gun batteries then under construction could command its length for some distance. For added security the Pennsylvania committee of safety appointed ten special Delaware pilots. Outside of Smith and the committee members, these men alone knew the exact location of the openings left in the chevaux-de-frise for friendly ships. The pilots were watched carefully and were restricted in their movements. They could be jailed indefinitely by the committee to protect against their falling into British hands.[24]

By September 29, seventeen chevaux-de-frise had been wrestled into

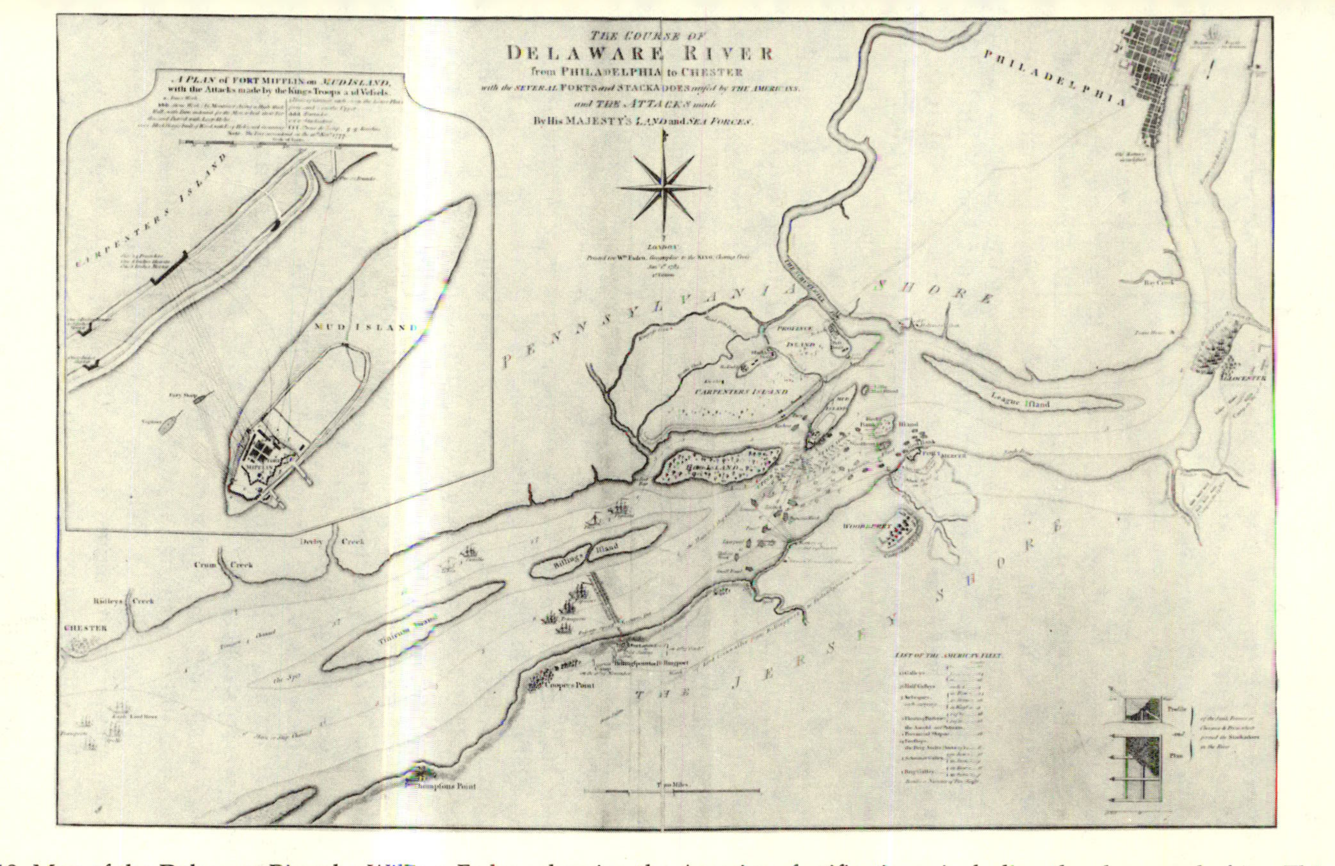

10. Map of the Delaware River by William Faden, showing the American fortifications, including the chevaux-de-frise. This map, drawn in 1785, is the third and presumably most accurate version of a map Faden first drew in 1778. Even then, Faden guessed at the exact location and arrangement of the chevaux-de-frise. Courtesy of the Historical Society of Pennsylvania.

place, at a cost of approximately £100 each. More were being built, but carpenters often had to wait for fresh supplies of logs to be gathered and transported to their Gloucester job site. Sixty-seven were eventually made. Members of the Continental Congress and the Philadelphia Assembly boarded some newly built galleys and toured the defenses and seemed to be satisfied with what Smith had done.[25]

Meanwhile other steps had been taken to fortify the Delaware. A plan tendered by Ephraim Anderson under the pseudonym "Nauticus" recommended the deployment of fireships. Consequently, firerafts heaped with straw, turpentine, barrels of oil, pine shavings, and other flammables were readied. These rafts lay berthed near the mouth of the Schuylkill River, opposite Mud Island. Eight iron chains from 250 to 280 feet long made at local forges joined the rafts together, six to a chain. The Pennsylvania state navy also bolstered its strength, with a brig on the quays at Philadelphia nearing completion. It eventually joined the fleet of thirteen galleys armed with one or two cannons apiece that had been assembled to guard the chevaux-de-frise. Other vessels, from schooners to small guard boats, were also added to the force. That this small navy could be built so quickly is testimony to the skill of Philadelphia shipwrights.

Naturally John Macpherson, having been turned away from Boston by George Washington, could not resist making some recommendations for improving this flotilla being assembled so close to his home. He advised the committee of safety to add keels and rounded hulls to the galleys for stability, and lateen sails for maneuverability. Philadelphia mechanic Nathan Irish also had some thoughts on the subject and presented the committee with a model of his proposed "Gundalo" or heavily armed galley. The recommendations of both Macpherson and Irish impressed the committee and may have been incorporated in some boats.

Finally, the patriots repaired and extended the incomplete earth and stone breastworks begun by the British a decade before on Mud Island. John Bull, a local militia officer, laid out a new redoubt across the river at Red Bank, New Jersey.[26] The earthworks at Red Bank were protected by a ditch and abatis on the land side and eventually mounted fourteen cannon of various calibers. Iron foundries in and around Philadelphia hurriedly cast cannon and shot for the two forts.

One other supplement to the Delaware defenses was proffered though not adopted. This was Joseph Belton's "submarine plan." Perhaps because it felt secure for the moment, lacked the requisite funds, or was skeptical of the whole scheme, the committee of safety chose not to act on Belton's petition.[27]

The one major flaw in the defense system blueprint went undiscovered. The "engineers" did not close the open passage between Hog Island and the west bank of the Delaware. After surveying the channel, they concluded that its entrance was too shallow to admit a warship and did not need to be blocked. Their decision not to cut off the one route circumventing the chevaux-de-frise opposite Mud Island cost them dearly two years later.

Nevertheless, Pennsylvanians had good reason to feel confident in their untried but formidable looking defenses. A substantial amount of work had been done in a comparatively short period of time and at their own initiative and expense. The forts were far from complete, but Philadelphia did not appear to be in any immediate danger of a full-scale, prolonged attack. The chevaux-de-frise alone were thought to be menacing enough to deter small-scale actions by the British navy. John Adams wrote his wife, Abigail, that he was so impressed by the chevaux-de-frise, firerafts, galleys, floating batteries, and gun emplacements that he had "no Suspicions of an Enemy from Sea."[28] Young Solomon Drowne, then attending lectures at the College of Philadelphia medical school, assured his parents back in Providence that he saw "no cause of Fear."[29]

This widely shared sense of security persisted into the spring of 1776. The Pennsylvania patriots had no foreknowledge of what lay ahead, so they went about their fort building at a leisurely pace after the initial war scare and surge of activity in the summer of 1775. They were almost too content. At this juncture a local citizen prompted them to upgrade their defenses. That sage advice came from amateur mathematician Daniel Joy in a report to the committee of safety on January 16, 1776. First, Joy recommended that the firerafts be improved. He precisely diagrammed their proper dimensions, the best types of combustible agents, and the most efficient method of lighting them. Next, he suggested that a chain be stretched between piers to shore up the gap between Mud Island and the chevaux-de-frise. In Joy's opinion the fort on Mud Island could not cover a large enough field of fire. "Ships of war will pay very little regards to Forts," he noted, "unless there be something besides to stop them." The whole network, he stated, should be activated by an early warning system consisting of beacons (buckets of tar suspended from poles) spaced every few miles to the Delaware capes. He also had some ideas on the batteries at Mud Island, pointing out that the use of sod or clay for the rear of those emplacements was preferable to stone, which caused cannon shot to ricochet. Joy's scheme to forge copper basins filled with boiling water for spraying British gun crews was far-fetched, but the reasoning behind his scheme was sound. "Look back into History," he contended, "and you'll find the New improvers in the art of War has always had the advantage of their Enemys."[30]

Acting in response to Joy's report, the committee of safety resolved that a chain boom be made to close the gap between the chevaux-de-frise and Mud Island. Ironmaster Robert Coleman of Lancaster County forged the links.[31] Solomon Drowne took time from his studies to see

Coleman's completed chain, "each Link of which is so heavy that I could scarce lift one of them from the ground."[32] Philadelphia carpenter Thomas Davis received a contract for building the log piers to which the chain would be fastened. Arthur Donaldson, who had been assisting Robert Smith in Gloucester, held the original contract, but in June 1776 he went to New York to instruct carpenters there on the making of chevaux-de-frise.[33] Workers sank and packed the two piers with stones in September. The chain was put in place two months later.[34] In addition, Joy's suggestion for an early warning system was adopted in modified form, with a line of signal guns teamed with dispatch riders serving the purpose even better than his proposed tar bucket beacons.

The final additions to the Delaware defenses prior to the 1777 campaign were a new line of chevaux-de-frise and a half-finished breastwork at Billingsport, two miles downriver from the first line of underwater obstacles. Excavation began at Billingsport late in 1776 because a growing number of patriots lost their complacency and feared that the defenses already erected might prove insufficient in the face of a full-fledged invasion. Indeed, with the reverses suffered by the American army that year in New York and Canada, invasion loomed as a distinct possibility.

Robert Smith supervised the building of the chevaux-de-frise sunk at Billingsport. Arthur Donaldson succeeded him after his untimely death in February 1777, Donaldson having returned from his New York assignment. Smith had seen to the completion of nearly all of the obstacles needed there, so Donaldson's primary job was to fill holes in those lines as new soundings were taken. Surplus chevaux-de-frise were moored at Mud Island, ready for immediate sinking in the lane left open for friendly ships should the British appear unexpectedly.[35]

At the same time these last-minute details were being attended to, the identity of the Delaware engineers began to change. European military engineers enlisted by agents of the Continental Congress finally slipped into the country during the summer of 1777 (Congress had been seeking them since December 1775). Where previously only local craftsmen such as Robert Smith and Arthur Donaldson labored, there were now professionals with impressive credentials. Pennsylvania had repeatedly sought such experts, but until the Europeans arrived, the Continental army had been unable to spare a single individual from its tiny cadre of engineers.[36] With the loss of New York and Washington's retreat into Pennsylvania, Pennsylvania's amateur engineers no longer worked in isolation. After the arrival of the professional engineers with Washington, these "amateurs" were still relied on, but for skilled labor, not strategic planning.

Although Thaddeus Kosciuszko was the first European military engineer involved with Delaware defense, he did little on the project and arrived before the main army under Washington. Kosciuszko had come to Pennsylvania from Poland in August 1776 and by the end of that year had been awarded a Continental commission. He inspected the fortifications taking shape at Red Bank and worked for a time on those proposed for Billingsport before being transferred to the northern army under Horatio Gates.[37] Pennsylvanians thus had had a trained engineer in their midst before 1777, but not long enough for him to accomplish much.

Phillipe Charles Tronson du Coudray, debarking from France in the spring of 1777 with a retinue that included eight engineers, became the first European to write a full report on the Delaware defenses.[38] In this critique he made recommendations running counter to what Kosciuszko had approved. For example, he regarded Billingsport as a strategic point but dismissed the fort Kosciuszko helped mark out there as practically useless for protecting the chevaux-de-frise, the key to the position. On the other hand, he believed the fort at Red Bank staked out by Colonel John Bull to be well designed, with "no other assistance than natural genius." Coudray hinted that the dimensions of the fort were perhaps too ambitious, but he assumed it would be garrisoned by a sufficient number of troops and did not press the point. Notwithstanding this, he concluded that Red Bank should be abandoned in favor of Billingsport, where the river was narrower and for that reason, in his opinion, more defensible.[39]

Coudray filed a second and more critical report in August. After touring the defenses with Arthur Donaldson, he concluded that the redoubt on Mud Island was underfortified and could be demolished in a few hours. Disregarding the notion that the channel between Hog Island and the Pennsylvania shore was too shallow for warships, he suggested that a new line of chevaux-de-frise, supported by land batteries opposite Mud Island, be emplaced to insure that the main line of defenses would not be outflanked. He reiterated his view that Red Bank ought to be evacuated and Billingsport be made the center of the defense.[40]

Coudray opted for a concentrated effort, while Kosciuszko had favored the multiple lines that Pennsylvania and New Jersey set up on their own initiative. Both Kosciuszko and Coudray endorsed the basic theory behind the combination of forts, ships, and chevaux-de-frise—the latter in particular—adopted by the Americans, even if they found deficiencies in application. Those deficiencies centered on the size of the fort at Red Bank, the relative strength and thickness of the walls both there and at Mud Island, and the concentration of effort.

Pennyslvania and the Continental Congress, coming to that state's aid after repeated complaints that it had exhausted its resources, chose to meet Coudray halfway. In deference to Coudray, Congress gave a new fort at Billingsport top priority, and construction began in the summer of 1777.[41] Yet it did not order the abandonment of the other forts, possibly because George Washington—in contrast to Coudray—wanted to keep the rest of the defensive works around Mud Island. Construction on forts Mifflin (Mud Island) and Mercer (Red Bank) continued. Beginning in October 1777, work at those posts proceeded under the direction of Francois Louis de Fleury and the Chevalier Mauduit du Plessis, respectively, both of whom were French military engineers.[42]

The new fort at Billingsport was barely more than a mound of dirt when Nathanael Greene reported enthusiastically on the defense preparations. "Upon the whole I would trust the security of the River to the Chevaux de Frizes, protected and defended by fort Mifflin, the shipping, and fire rafts." Given the narrowness of the navigable channel below Mud Island and the sturdiness of the underwater obstructions, Greene claimed, "There is very little danger of the enemys opening themselves a passage."[43] Washington also surveyed the defenses but arrived at a more realistic conclusion. He saw that the Delaware could not be held forever against a combined British land and naval assault. If it could just be held until winter, when the river froze, any British move toward Philadelphia would be hampered by the difficulty of overland supply. In Washington's opinion, at the very least the unfinished redoubts at forts Mifflin and Mercer had to be finished. Together with the earthworks at Billingsport, they were the weakest links in the Delaware defense chain. He also agreed with Coudray that the passage between Hog Island and the Pennsylvania shore was vulnerable and needed a line of chevaux-de-frise.[44] Unfortunately, events moved too quickly, and the British sailed up the Delaware before the new chevaux-de-frise could be assembled and the forts completed.

A British expeditionary force under General Sir William Howe had left Staten Island on July 23, 1777. Learning that the American army was elsewhere and that the Delaware appeared to be heavily defended, Howe thought better of landing at the mouth of that river and sailed into the Chesapeake. After landing at the Head of Elk, he marched his army northeast, drove off Washington at the Brandywine River on September 11, and seized Philadelphia fifteen days later. Howe's next logical move was to secure his line of supply and communications before the Delaware became choked with ice; otherwise he could not remain in Philadelphia through the winter. Wagon trains carrying precious supplies were too susceptible to ambush by Washington, whose army had been stung but not beaten. So Lord Richard Howe, brother of the general, began naval operations to clear the Delaware in October. If the combined army and navy contingents under the two Howes could not clear the Delaware, the success of the entire British campaign was in jeopardy.

British warships making their way upriver experienced considerable difficulty in forcing a corridor through the chevaux-de-frise at Billingsport. The unfinished fort there had been hastily abandoned without a fight on October 2, but Pennsylvania state navy galleys sallied forth and harassed British crews grappling with the underwater obstacles. After back-breaking labor, British crews moved two frames, and small frigates squeezed through on the nineteenth. Without support from land batteries, the galleys and chevaux-de-frise could not stop them, a possibility that Nathanael Greene and others had not fully appreciated. While the British naval column prepared to move against the staggered lines of chevaux-de-frise opposite Mud Island, a land force assaulted Red Bank, only to be bloodily repulsed. Fort Mifflin held on until November 15, when the garrison evacuated it after the walls had been pounded to rubble by shore batteries on Province Island and the *Vigilant*, a converted merchantman. The *Vigilant* had crossed into the back channel on the Pennsylvania side of Hog Island, bypassing the chevaux-de-frise guarding the main channel. The fate of Fort Mercer (Red Bank) was sealed with the fall of Fort Mifflin. The garrison abandoned it, and the remaining ships of the Pennsylvania state navy were scuttled, seized, or withdrawn far upriver. By the end of November the British had the Delaware firmly in their grasp.[45]

After the American defeat, there followed the usual recriminations, charges, and countercharges. M. Fleury, engineer at Mud Island, accused the Pennsylvania navy of failing to recognize its principal function of protecting the chevaux-de-frise.[46] Fleury, who had been on Mud Island only two weeks, and more specifically his predecessors, in turn came under fire for not doing more to reinforce the woefully underfortified north side of Fort Mifflin. Sir James Murray, a British army officer, viewed Fort Mifflin as a strong point that fell primarily because it was unfinished. "If the Rebels had taken every advantage of their situation," he confided to his wife, "it is not very certain that we should have been able to have cleared the passage at least this campaign."[47] The Marquis de Chastellux averred in his *Travels* that the key to the fall of Fort Mifflin had been the undefended channel between the Pennsylvania shore and Hog Island. His charge that Fort Mercer was too extensive and undergarrisoned seemed to be representative.[48] Still others emphasized that the Americans were doomed from the beginning because the British had already won the land campaign in Pennsylvania, and Washington was too weak to assist the Delaware forts.

In truth, failure resulted from combined military and engineering deficiencies. The American army had been outmanned, outgunned, and, at the Brandywine, outmaneuvered. The British army's subsequent command of movement in the countryside rendered precarious the existence of the Delaware forts. Furthermore, the Pennsylvania navy was no match for Lord Richard Howe's frigates, and American troops garrisoned along the Delaware were too few and unused to either land or water siege operations. Engineering breakdowns compounded an already bad situation. Because the forts at Red Bank, Mud Island, and Billingsport stood incomplete, the Delaware defense system proved to be flawed at crucial points, and the British exploited those flaws. Undeniably the fortifications on Mud Island were simply inadequate. Even if the back channel had been obstructed by a new wedge of chevaux-de-frise, the British batteries on Province Island alone may have been enough to silence American guns.[49] The defects of Fort Mercer can be attributed to both the small garrison and its overambitious layout.

When all of the foregoing are added together, it would seem that defeat was inevitable. It is understandable even if regrettable that failure to stop the British has obscured the real accomplishments made by local craftsmen working on the Delaware defenses. Robert Smith in particular deserves recognition. His chevaux-de-frise performed as intended, a fact attested to by the numerous enemy vessels sunk or severely damaged that happened onto them. Large warships attempting to detour around the chevaux-de-frise invariably ran aground—convincing evidence that those who took soundings did their job well.[50] Both the sixty-four-gun ship *Augusta* and the sloop *Merlin*, which came to the *Augusta*'s assistance, were destroyed after they became caught in the channel. Other British warships plied the waters very carefully to avoid a similar fate.[51] Indeed, Major John André noted soon after the river campaign's conclusion that "the weighing [of] the Chevaux de Frize was said to be a work of great Labor not to be effected." Verifying André's apprehensions, British vessels making for Philadelphia had to traverse the narrow opening left by the patriots for their own ships.[52] Subsequent attempts to clear a wider passage brought minimal success.

Pennsylvania's "engineers" felt that no sizable warship could negotiate the entrance into the channel behind Hog Island, so they left that avenue unobstructed. Their decision proved to be a costly, possibly even fatal, error, especially because of the unsatisfactory conditions at Fort Mifflin. This one oversight by the original planners hastened the collapse of the rest of the defenses. But in all fairness it should be noted that the *Vigilant* only gained entry into the passage behind Hog Island after being temporarily stripped of its guns to lessen the draft.[53] Even if Smith, Donaldson, Franklin, and the rest had foreseen such an eventuality, Mud Island may have been held a day or two longer at best. Overall success hinged on too many components, all of which had to work in perfect harmony. Blame, if blame is the right term, cannot be laid on any one doorstep.

The campaign of 1777 demonstrated graphically that the Delaware defenses were less than perfect. Indeed, some of those defenses were wholly inadequate. But inability to check the British stemmed in part from intangible, unforeseen factors. Devising an "impregnable" defense may have been impossible, given military realities. When all is said and done, the fact still remains that the British did not sweep up the Delaware. They took several weeks; more time, certainly, than they needed to pass the forts on the lower Hudson back in the summer of 1776. Moreover, the arrival of professional military engineers did not bring a panacea for all ills. Those engineers were not always of one mind, and they could not single-handedly overcome the problems associated with insufficient men, materials, and time. Plessis wisely reduced the size of Fort Mercer to make it more defensible, and Fleury did what he could at Fort Mifflin, but their improvements were stopgap measures at best.

Though on their own for nearly two years, Pennsylvania's "engineers" rose to the occasion. Acting without professional guidance, they erected an ingenious network of forts, underwater obstacles, and auxiliary defenses, some effective, some not so effective. In those areas where their practical expertise was most valuable—as in designing the chevaux-de-frise—they enjoyed striking success. Where they were untrained and inexperienced—as with the forts—they did not always do so well. Given their limitations, Pennsylvania craftsmen did a creditable job as interim engineers. The exigencies of local defense stimulated and fed on their application of new tools and innovative ideas.

Robert Smith's career as a wartime inventor was therefore quite different from that of David Bushnell. Bushnell had invented a device that no one had expected to see, and observers were not sure how to react. Smith invented devices that answered a specifically defined need. He did not have to persuade listeners that his chevaux-de-frise could be useful; they already knew that. Bushnell had to arrange his own financing; Smith did not. Smith's devices were clever but simple; Bushnell's "Turtle" required a greater imaginative leap. Delaware defense had provided an inventive outlet for the skills of craftsmen like Robert Smith, and the same could be said of other engineering projects. Ingenuity did not disappear during the war. Americans after the war would point with pride to the accomplishments of their craftsmen as they began to notice these inventive geniuses in their midst. They may not have remembered the Bushnells or Beltons, or attached much significance to submarine experiments, but they did remember what craftsmen on the Delaware and elsewhere had done.

NOTES

1. Burnett, ed., *Letters*, I, 169-170. Even Great Britain lacked a community of civil engineers. As stated by Paul Mantoux: "their place was more or less filled by carpenters, locksmiths, or clock makers, in fact by anyone who was sufficiently used to working in wood or metal, and who could set up wheelwork or fit parts of machinery together"; see Mantoux, *The Industrial Revolution*, p. 216. Harrison statement in Burnett, ed., *Letters*, I, 169-170.

2. Greene to General Joseph Varnum, letter of 1 August 1777, in the Nathanael Greene Papers, Henry E. Huntington Library, MSS.

3. Hazard, ed., *Pa. Archives*, 8th series, VIII, 7239.

4. Ibid., 2nd series, I, 753.

5. Hazard, ed., *Pa. Col. Records*, X, 280, 282.

6. See Bridenbaugh, *Colonial Craftsman*, passim; and idem, *Rebels and Gentlemen*, passim, for background material.

7. Hazard, ed., *Pa. Archives*, 1st series, IV, 635-636; and idem, ed., *Pa. Col. Records*, X, 284.

8. Hazard, ed., *Pa. Archives*, 8th series, VII, 7239.

9. Ibid., 2nd series, I, 752-754. Wells presented sketches of two other devices, but they were not accompanied by written descriptions. His pump for leaky vessels is diagramed and described in the *APS Trans*. III (1771):289-292; see chapter 2 for further information; and James T. Flexner, *Steamboats Come True* (New York: Viking Press, 1944), pp. 93-95, 385, for Wells's and Donaldson's roles in the Fitch-Rumsey steamboat controversy. For Robinson see Hazard, ed., *Pa. Col. Records*, X, 286; and idem, ed., *Pa. Archives*, 8th series, VIII, 7171, 7178, for his pump. Though his plan was rejected, Robinson later contracted to build machines designed by Robert Smith (see ibid., 2nd series, I, 755-756). For a sketch of Donaldson's inventive career before the war see chapter 2.

10. For details on Smith's life, see C. P. Stacey, "Robert Smith," in Malone, ed., *Dict. Amer. Bio*., XVII, 335-356. Joseph Harris's sketch of Smith in the *PMHB* 4 (1880):79-88 is unreliable because the author confused the lives of Smith (1722-1777) and a Colonel Robert Smith (1720-1803), both of whom were residents of Chester County.

11. Joseph Jackson, *Early Philadelphia Architects and Engineers* (Philadelphia: Joseph Jackson, 1923), p. 69.

12. Charles E. Peterson, "Carpenters' Hall," in Luther P. Eisenhart, ed., "Historic Philadelphia," *APS Trans*. 43 (1953), part 1, pp. 120-123; Hazard, ed., *Pa. Archives*, 8th Series, VII, 6335-6336; VIII, 7181, 7204-7205, 8187, for the Schuylkill River bridge; and Labaree, ed., *Papers of Franklin*, X, 237; XII, 168, for Smith and Franklin's house.

13. *Pennsylvania Packet*, 15 and 27 June 1774.

14. Hazard, ed., *Pa. Col. Records*, X, 290.

15. Carl Van Doren, *Benjamin Franklin* (New York: Viking Press, 1938), p. 534; also the letter from Franklin to Charles Thomson of 29 December 1788 in Smyth, ed., *Writings of Franklin*, IX, 696. Henry Bellas, "The Defenses of the Delaware River in the Revolution," *Wyoming Historical and Geological Society Proceedings and Collections* 5 (1900):51, attributed the devices to Franklin. Labaree, ed., *Papers of Franklin*, XXII, 73n, notes the debate and includes excerpts from some autobiographical notes Franklin made in May 1786, where he appeared to take credit. Boyd, ed., *Papers of Jefferson*, IX, 497, 497n, points out that these notes were dictated, not written by Franklin himself. Perhaps Franklin's memory had dimmed or his comments were garbled. John W. Jackson, *The Pennsylvania Navy, 1775-1781* (New Brunswick: Rutgers University Press, 1974), pp. 354, 356, reached conclusions similar to my own.

16. Called stackadoes on a map of the Delaware engraved by William Faden

of London in 1778 and reproduced in Leonard Lundin, *Cockpit of the Revolution* (Princeton: Princeton University Press, 1940); bugbear reference in *Naval Documents*, IV, 1314; and caisson in ibid., III, 941.

17. Cited in Wallace McGeorge, "The Chevaux-De-Frise in the Delaware," a paper read before the Gloucester County Historical Society on 18 July 1911. McGeorge included many valuable documents in his account, but some of his conclusions are erroneous.

18. At least a score of descriptions of the chevaux-de-frise have survived. The most complete are in Roger Lamb, *Journal of Occurances During the Late American War* (Dublin: Wilkinson and Courtney, 1809), p. 232, which may or may not have been extrapolated from David Ramsay, *The History of the American Revolution*, 2 vols. (London: J. Johnson, 1791), II, 17. Ramsay's history was published before Lamb's journal, but Lamb was an eyewitness, and Ramsay may have seen Lamb's manuscript. Also see William Gordon, *The History of the Rise, Progress and Establishment of the Independence of the United States of America*, 4 vols. (London: William Gordon, 1788), II, 93; an anonymous "Contemporary British Account," *American Antiquarian Society Proceedings*, new series, 40 (1930):83-84; Charles Stedman, *The History of the Origins, Progress, and Termination of the American War*, 2 vols. (Dublin: P. Wogar et. al., 1794), I, 296, 331; and John Adams's note in a letter of 31 October 1775 in the GWP, series 4 (reel 34). The number of logs used can be found in Hazard, ed., *Pa. Archives*, 1st series, IV, 774. A convenient collection of chevaux-de-frise documents, taken primarily from the *Pa. Archives* is in Jackson, *The Pennsylvania Navy*, appendix B, pp. 353-376. Also see Samuel S. Smith, *Fight for the Delaware: 1777* (Monmouth Beach: Philip Freneau Press, 1970), pp. 8-9, and passim.

19. Lamb, *Journal*, p. 232.

20. Gordon, *History*, II, 93.

21. Hazard, ed., *Pa. Archives*, 1st series, V, 66.

22. Idem, *Pa. Col. Records*, X, 299. The committee had planned to chain the chevaux-de-frise together to frustrate any attempt to move them, but since the devices were sunk in small lots or individually, it seems unlikely that Smith carried out this part of the plan.

23. Ibid., X, 288, 289, 294; Hazard, ed., *Pa. Archives*, 1st series, IV, 785-786. The location of the chevaux-de-frise is mentioned in Nicholas Cresswell, *Journal, 1774-1777* (New York: Dial Press, 1924), p. 154. David Rittenhouse did not become the Pennsylvania committee of safety engineer until 27 October 1775; hence he probably did not participate in the site selection.

24. Hazard, ed., *Pa. Col. Records*, X, 363; idem, *Pa. Archives*, 1st series, IV, 777-778; 2nd series, I, 359; 4th series, III, 589-591.

25. Hazard, ed., *Pa. Col. Records*, X, 334, 344, 350-351; and Smith, ed., *Letters of Delegates*, II, 76, 78, for the congressional excursion. The chevaux-de-frise accounted for £1700 out of £87,237 in defense expenditures, most of which went toward the purchase of munitions in Europe. Two receipts, dated 6 April and 12 June 1776, made out to Samuel Wheeler for ironwork on the chevaux-de-frise, are located in the Gratz Collection, Pennsylvania Series, Navy Board, Case 1, Box 18, HSP MSS. (under Samuel Morris, Jr.). This was probably the same Wheeler mentioned in the preceding chapter, in note 71.

26. Galleys in Jackson, *Pennsylvania Navy*, p. 19; Macpherson in *Naval Documents*, II, 240; Irish in Hazard, ed., *Pa. Archives*,1st series, IV, 648; also see ibid., IV, 666-667; and idem, *Pa. Col. Records*, X, 366.

27. Hazard, ed., *Pa. Archives*, 1st series, IV, 650-652, 654.

28. *Naval Documents*, V, 1290.

29. Harold E. Gillingham, ed., "Doctor Solomon Drowne," *PMHB* 48 (1924):242; for an enthusiastic report similar to Drowne's, see *Naval Documents*, II, 1307-1308.

30. Hazard, ed., *Pa. Archives*, 1st series, IV, 698-701.

31. The appearance of the British frigate *Roebuck* near the mouth of the Delaware in March 1776 accelerated the concern. For Coleman see Joseph R. Delafield, "Notes on the Life of Robert Coleman," *PMHB* 36 (1912):226-230; and Hazard, ed., *Pa. Archives*, 1st series, V, 39.

32. Gillingham, ed., "Doctor Solomon Drowne," p. 246.

33. Hazard, ed., *Pa. Col. Records*, X, 525, 530, 648, 717, 721, 783; Donaldson to the Hudson in idem, ed., *Pa. Archives*, 1st series, IV, 774. For a description of the chevaux-de-frise made there, see Heath, *Memoirs*, p. 40. Captain John Hazelwood went along to give directions on building fire rafts.

34. Hazard, ed., *Pa. Col. Records*, X, 717, 726, 783.

35. For the condition of the Delaware defenses in May 1776 refer to ibid., X, 575-578; for Smith at Billingsport see Hazard, ed., *Pa. Archives*, 1st series, IV, 784; V, 8, 9, 37, 283; Donaldson in ibid., 2nd series, I, 93, 103-104, 167; chevaux-de-frise moored at Mud Island in ibid., 1st series, V, 45; notice of Smith's death in the *Pennsylvania Evening Post*, 13 February 1777. In August 1776 Robert Erskine, who had worked on the Hudson defenses and later gained fame as George Washington's mapmaker, sent a design of his "Tetrahedron"—a simple wooden frame intended to function in the same manner as Smith's chevaux-de-frise—to Benjamin Franklin. Franklin wrote back that the machine was not practical for the soft bed of the Delaware, and Smith's devices were more than satisfactory. See the Franklin Papers, IV, fos. 102 1/4 and 102 3/4 for Erskine's letter of 16 August 1776 and Franklin's answer of 16 October 1776 (along with Erskine's drawing of his "Tetrahedron") in the APS MSS.

36. Hazard, ed., *Pa. Archives*, 1st series, IV, 775; Idem, *Pa. Col. Records*, X, 604.

37. Miecislaus Haiman, *Kosciuszko in the American Revolution* (New York: Polish Institute of Arts and Science, 1943), pp. 1-12.

38. The names of the engineers with Coudray are listed in the PCC, Item 156 (reel 176), among Coudray's papers.

39. Coudray's report is reprinted in Hazard, ed., *Pa. Archives*, 1st series, V, 360-363. See Paul K. Walker, *Engineers of Independence* (Washington, D.C.: U.S. Army Corps of Engineers, 1981), pp. 146-174, for a discussion of Coudray and Washington, with a number of documents included.

40. Report in the GWP, series 4 (reel 43), letter to Washington of 30 August 1777.

41. See the report of the Pennsylvania Supreme Executive Council to Washington of 18 July 1777 in Hazard, ed., *Pa. Archives*, 1st series, V, 431-432. The overall defense layout is summarized succinctly in Lundin, *Cockpit of the Revolution*, pp. 336-337.

42. Reviewing the fortifications at Red Bank in 1781 the Marquis de Chastellux followed Plessis in noting that the Americans "little practiced in the art of fortifications" had laid out works "beyond their strength to hold." See Chastellux, *Travels in North America*, 2 vols. (Chapel Hill: University of North Carolina Press, 1963), I, 157.

43. Cited in Worthington C. Ford, ed., *Defenses of Philadelphia in 1777* (Brooklyn: History Printing Club, 1897), p. 5.

44. Fitzpatrick, ed., *Writings of Washington*, IX, 45-53. Washington made his report in August. Back in June, Benedict Arnold, who was commanding the militia stationed along the river, warned that the defenses were inadequate. In addition to criticizing the redoubt at Billingsport, he noted that Mud Island could be shot at from the west bank. See his report to Congress of 10 June 1777 in PCC, Item 162, I, 90-91 (reel 179).

45. See the indispensable collection of documents in Ford, ed., *Defenses of Philadelphia*, passim; and the narratives in Lundin, *Cockpit of the Revolution*, pp. 336-371; Jackson, *Pennsylvania Navy*, pp. 120-281; Christopher Ward, *The War of the Revolution*, 2 vols. (New York: Macmillan Co., 1952), I, 373-383; and Ira D. Gruber, *The Howe Brothers and the American Revolution* (Chapel Hill: University of North Carolina Press, 1972), pp. 246-260.

46. Jackson, *Pennsylvania Navy*, p. 161.

47. Letter from Murray to his wife of 29 November 1777 in Eric Robson, ed., *Letters From America, 1773-1780* (Manchester: University of Manchester Press, 1951), p. 50.

48. Chastellux, *Travels*, I, 154-157; also Thomas Paine's letter to Benjamin Franklin of 16 May 1778 in *PMHB* 2 (1878):292.

49. For the Province Island batteries see G. D. Scull, ed., "Journals of Capt. John Montresor, 1757-1778," *New York Historical Society Collections*, XIV, 462-477. Montresor was the engineer who laid out the original fort on Mud Island in 1771.

50. Major John André, Journal, 1777-1778, Huntington Library MSS., p. 52. Arthur Donaldson and Levi Hollingsworth were hired by the Port Wardens of Philadelphia to remove the chevaux-de-frise in 1784. Some of the sixty-seven had been moved by currents and ice, but all except the three that had been removed by the British were still in the river, though some had decayed badly. See Hazard, ed., *Pa. Archives*, 2nd series, I, 758-769.

51. The *Augusta* may have simply run aground, though one American claimed that it had "got onto the chevaux-de-frise." See Joseph Plumb Martin's narrative, edited by George Scheer as *Private Yankee Doodle* (Boston: Little, Brown and Co., 1962), p. 87. The *Merlin* apparently impaled itself on one of the devices. Both ships were later completely destroyed by gunfire.

52. André, Journal, Huntington Library MSS., p. 58.

53. Letter from Nathanael Greene to George Washington of 14 November 1777, in the Greene Papers, Huntington Library MSS.

6 LIMITS TO INNOVATION: THE PENNSYLVANIA RIFLE

At the beginning of the war, leading Americans confidently expected the Pennsylvania rifle, one of the most notably innovative tools of the Colonial Era, to help them secure a swift victory. This firearm was to become, temporarily at least, a source of fierce national pride. Many assumed that it would give the American soldier a ready-made advantage over his musket-toting British counterpart. The Pennsylvania rifle, with its peculiar characteristics adapted to and evolved from the frontier experience, stood then and continues to stand as a monument to colonial ingenuity. As Roger Burlingame penned effusively, the rifle "was the truest kind of American invention, the certain product of an American culture."[1]

Popularizers of the rifle have been numerous and vocal. They paint a picture of colonial riflemen—rough-hewn giants of the primordial forest—marching forth to vanquish the minions of George III in more or less the same manner they furthered the course of westward empire. What riflemen did in the woodlands to advance civilization they did on the seaboard to ward off defeat from behind, or so the story goes. Predictably enough, when writing about the rifle's contribution to the war, Burlingame concluded that "it was largely this terror which brought the Revolution to an end victorious to America."[2] Burlingame's filiopietistic notions about the rifle have been shattered, and the myth of the rifle's omnipotence has been all but dispelled. Still, there is a lingering mystique surrounding the rifle, perhaps because the rifle is so often thought of in connection with American genius.

The rifle was an important colonial innovation, but it was by no means a super weapon, and it was not used very effectively during the War of Independence. Wartime use of the rifle showed that a new weapon is

---

Portions of this chapter originally appeared as Neil L. York, "Pennsylvania Rifle: Revolutionary Weapon in a Conventional War?" *Pennsylvania Magazine of History and Biography* 103 (1979): 302-324.

ineffective if there is not a new doctrine to go along with it and an industrial sector capable of producing sufficient quantities. American political and military leaders going into the conflict did not know how the rifle could best be used or if its use would prove more disruptive than productive. But they could not resist trying to capitalize on its supposed superiority. They went ahead and used rifles; many later regretted having made that choice. Rifles were expensive, hard to make, and difficult to use. The munitions industry could not make weapons of any sort fast enough, and there was no simple way to accelerate production. The rebellious colonies lacked the managerial experience with large-scale enterprises and centralized governmental bureaucracy to build a munitions industry around the rifle. They had neither the capital nor the workers to expand their shops or their methods of production. Furthermore, those who wanted to use the rifle were either unable or unwilling to reorganize the American military around any single weapon, however superior it might have been—and in some ways the rifle was a superior weapon.

The superiority of rifled gun barrels had first been discovered in the late fifteenth or early sixteenth centuries by gunsmiths in central Europe. Whether by accident or experiment, they found that a gun barrel scored with spiraling lands (high points) and grooves (depressions) was much more accurate than a smoothbore weapon. Spinning motion imparted to a bullet made it fly truer, cutting down on loss of velocity and propensity to windage.[3] Hunters using rifled weapons shot game with greater ease than ever before. Rifles consequently spread from the forests of Germany to other areas of Europe.[4]

According to most accounts, large numbers of rifles were first brought to the American colonies by German immigrants around 1700. Pennsylvania became a center of the trade, and numerous Pennsylvania gunsmiths earned a comfortable living by specializing in riflemaking.[5] By 1750 rifles were common in frontier communities along the length of the Alleghenies, for Pennsylvania craftsmen had never completely monopolized the art of rifling. Just prior to the War of Independence shops had spread to Baltimore, Maryland; Alexandria, Cumberland, Winchester, and Richmond, Virginia; Camden, South Carolina; Salisbury and Augusta, Georgia; and a few Pennsylvania gunsmiths had reportedly migrated to western New York.[6]

The rifle went through a metamorphosis in Colonial America and differed strikingly from its European forebear. A few years after importation it became obvious that modifications were desirable if not absolutely necessary to better adapt the rifle to the frontier. Colonial gunsmiths thus made basic alterations in rifle construction leading to a distinctive American archetype.

American backwoodsmen found that the short and heavy European

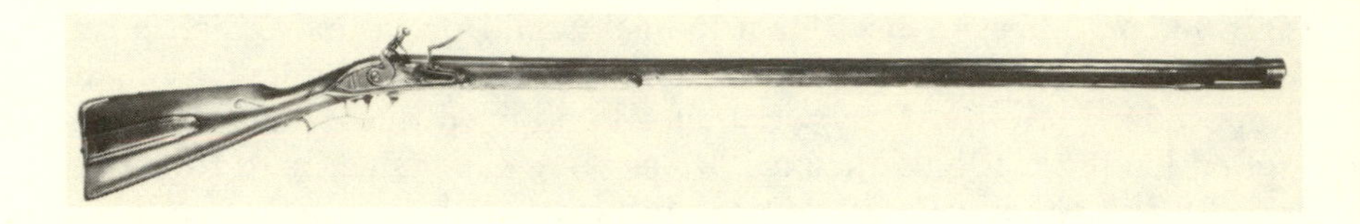

11. A Pennsylvania rifle, ca. 1760. Smithsonian Institution Photo No. 79-7928.

rifles that had evolved by the 1720s were incompatible with their needs. Since American backwoodsmen hunted out of necessity, not love of sport, they wanted an ever more accurate weapon. Gunsmiths accordingly lengthened the barrel to increase accuracy. In addition, they reduced the caliber and exterior barrel dimensions to cut down on weight. Backwoodsmen who tramped through the forests for weeks at a time appreciated this last modification in particular. Greased patch and hickory ramrod totally eclipsed the mallet and iron ramrod of earlier years because the patch-wrapped bullet took less time to load and helped guard against the accumulation of fouled powder in the barrel. Quick repetition of fire was indispensable for hunting and was even more essential for the hit-and-run tactics of Indian warfare. Such warfare was a fact of frontier life, and those living on the frontier needed a dependable weapon whose barrel did not require constant swabbing. Colonial riflemakers also made dozens of minor alterations, from casting thicker trigger guards to selecting choice native hardwoods like curly maple and cherry for gunstocks.

These improvements enhanced the rifle's reputation as a precision firearm. By 1760 it had become a true work of art as well as a useful tool, a harmonious blending of iron, brass, and wood. It outstripped the smoothbore musket in accuracy and sophistication of design. Yet the musket was by far the most commonly used weapon during the War of Independence and for half a century after. On first observation this seems inexplicable. Muskets were accurate up to a range of eighty yards; rifles were deadly at thrice that distance. Muskets generally had a larger bore, but what the rifle surrendered in knockdown power, it more than made up for in ease of carrying. Pound for pound a rifle was more efficient, at least for hunting and backwoods skirmishing.

At the same time, the musket had certain advantages over the rifle as a military weapon. First, it could use coarser powder. Cleaning fouled powder out of a grooved barrel was difficult, and riflemen had to choose their powder carefully. Soldiers armed with muskets did not have to be quite so careful. Besides, American troops during the War of Independence were lucky to have any gunpowder at all. Improved refining to accommodate rifles would have been prohibitively expensive. Second, most rifles were made according to the users' specifications, not a uniform design. There were as many different rifles as there were riflemen. Riflemen therefore prepared their own cartridges, usually on the spot. Some of them also owned weapons with elongated barrels and narrow, delicate stocks which, while aesthetically pleasing, made their guns impossible to modify. Muskets were more "standardized"—the term is applied loosely here. Musket-equipped soldiers often fired prepared cartridges for guns that could be assembled and replaced more cheaply and in less time. Third, troops armed with muskets could ordinarily load more quickly since they did not have to worry about a snug fit. While the rifleman carefully tamped his patch-wrapped ball down the barrel, the musket-user could pour ball and powder down the barrel in one easy stroke, a movement of reflex rather than skill. Fourth, muskets made for the military generally came with bayonets while rifles did not. That advantage was tied to yet another: muskets better served the purposes the military had in mind.

Muskets were admirably suited to the volley fire tactics of the day. After two or three exchanges of gunfire at close range—perhaps fifty or sixty yards—European military thinkers believed that bayonets should decide the battle, so their manual of arms dealt only superficially with the proper aiming and firing of muskets. Many European tacticians viewed muskets as merely convenient handles for bayonets, and some Americans felt the same. European foot soldiers were trained to fight in a solid line arrayed in an open field, ready at any moment for the tide-turning thrust of "cold steel."[7] So too, eventually, were most American regulars. The Continental army under George Washington ended up being a slightly modified version of European armies. Small numbers of riflemen served alongside as auxiliaries responsible only for skirmishing and scouting. The individualistic type of warfare practiced by irregulars in Europe and riflemen in America "did not fit into the eighteenth century European pattern, and European habits died hard."[8]

In a war where the enemy preferred to follow traditional precepts, backwoods riflemen were confronted by their antithesis. American riflemen fought a mobile style of war, putting a premium on expert shooting, not massed volley fire; concealment, not open field formation; and quick movement, not the measured cadence of a linear assault. American rifles lacked bayonets, since bayonets represented a different martial philosophy, a philosophy of limited worth in the forest. Most rifles could have been adapted to take bayonets, but that did not mean the riflemen's performance against British regulars would have improved correspondingly. Differences in fighting techniques went much deeper. Musket and rifle symbolized antipodal approaches to war. The American rifle had evolved from a different set of demands, a different mode of life. Frontier riflemen during the War of Independence were expected to learn how to fight differently. Adjusting to the change unnerved many and proved impossible for others.

Neither the riflemen nor their admirers knew this at the outset of the war. Realization came only with time, so when need arose for volunteers to assist New Englanders laying siege to Boston in the summer of 1775, patriot leaders turned to riflemen. George Washington thought they would make excellent soldiers. Remembering his harrowing experiences with Braddock and the limitations of regulars in forest warfare, he looked to independent-minded riflemen, expecting them to form the core of the Continental army. Numerous congressional delegates shared Washington's faith. Richard Henry Lee boasted that Fincastle County, Virginia, and five other western counties could raise one thousand riflemen each, all of whom could hit an orange from two hundred yards.[9] John Hancock had yet to actually meet a rifleman, but the guarantees of Pennsylvania, Maryland, and Virginia leaders made him their champion. "They are the finest marksmen in the world," he exclaimed, "they do execution of their Rifle Guns at an amazing distance."[10]

Americans like Washington, Lee, and Hancock were proud of their riflemen. Riflemen helped them to develop a national sense of identity. Throughout the colonial period Americans had been conscious of their cultural inferiority to the mother country. They came to resent that inferiority, just as they came to resent their place in the empire. The rifle symbolized the same thing to them as home manufactures: evidence of achievement, and therefore evidence that they were not innately inferior. Whatever inferiority they might admit to, they blamed on the politics and economics of empire. The rifle assured them that they were an ingenious people. In celebrating their perceived ingenuity, they did not celebrate the rifle as an invention per se, however. The editors of the *Pennsylvania Gazette* had not even included it in their 1771 list of "American inventions." Craftsmen who designed and assembled rifles were not singled out for special notice even though the frontiersmen who used rifles were. William Henry, an innovative Lancaster, Pennsylvania, gunsmith who made hundreds of rifles, is virtually forgotten. Daniel Morgan, a Virginian who used a rifle, is remembered as one of the great heroes of the War of Independence. He and his comrades in arms were legends in their own time. Their countrymen expected a great deal from them.[11]

Consequently, on June 14, 1775, Congress resolved that "six companies of expert riflemen, be immediately raised in Pennsylvania, two in Maryland, and two in Virgina."[12] Counties along the Susquehanna River in Pennsylvania seemed to be overrun with eager volunteers, so Congress amended that state's quota from six to eight companies, those eight companies to be formed into an independent Pennsylvania rifle regiment. Even then Lancaster County had too many volunteers, so it organized two companies, Congress gave its assent, and the rifle regiment went from eight companies to nine.[13] Meanwhile Virginia and Maryland had no difficulty in bringing their four companies to strength. Volunteers swarmed in as they had in Pennsylvania. The Virginians in particular were reputed to be fierce warriors, many having seen action in Lord Dunmore's War. In all, over 1,250 riflemen marched to Boston when Congress had originally called for slightly less than one thousand.

Congress made certain the companies were outfitted in grand style. Colonel William Thompson, commander of the Pennsylvanians, and the Pennsylvania state government were allocated $15,000 between them for the rifle regiment's expenses. At Reading and other towns along the route to Massachusetts, the men received new rifles, knapsacks, blankets, and canteens. Marylanders and Virginians enjoyed the same treatment.[14]

Riflemen caused a stir in each town they passed through. Newspapers noted the progress of their march to Boston anxiously, expectantly. A Baltimore resident reported that Daniel Morgan's company of Virginians looked "truly martial, their spirits amazingly elated, breathing nothing but a desire . . . to engage the enemies of American liberty."[15] The *Boston Gazette* noted the arrival of some riflemen in August, describing them as "an excellent Body of Troops . . . heartily disposed to prosecute, with the utmost Vigour, the Noble Cause in which they are engaged."[16] And the speed with which the companies made their trek boosted their reputation, the Pennsylvanians from Cumberland County covering 441 miles on foot in twenty-six days.[17]

Rifleman attire may have astounded some New Englanders. Most of the riflemen wore buckskin breeches, some with belts of wampum tied around the top. Almost all had on wool or linen hunting shirts, ranging from ash-colored to deep brown or dark grey. A few sported moccasins ornately decorated with porcupine quills. Tomahawk, hunting knife, soft felt hat, powder horn, and bullet pouch completed their garb.[18] Washington would have preferred that the whole army be so attired, not only because of lightness and natural camouflage, but also because it would help remove "those Provincial distinctions which lead to Jealousy and Dissatisfaction." In addition, Washington hoped the British would fear everyone so dressed as a deadly marksman.[19]

At Washington's behest the riflemen displayed their sharpshooting skills in Cambridge, as they had at various points along their line of march. They purportedly gave an extraordinary show of accuracy with their weapons, hitting a mark from 200 yards—some doing this while on the "quick advance." Others struck targets seven inches in diameter from a range of 250 yards with a precision no musket-carrying New Englander or British regular could hope to match.[20] Leaving their audience agape, they bivouacked in a special area and were exempted from routine duties. Washington, it would appear, wanted to put his model soldiers on display.

Riflemen initially caused a furor within the British lines. British sentries were not accustomed to being shot at so accurately. Catching their opponents napping, nefarious "rebel" sharpshooters picked off the careless and unwary by long-range shots or in twilight sorties. Patriot newspapers gleefully followed their exploits.[21] The propaganda value of the rifle aside, however, British soldiers soon adjusted to their new menace. They stayed safely out of sight behind their breastworks.

It did not take long for some American officers to discover that their great expectations for the riflemen were ill-founded, both as a result of the riflemen's temperament and the state of affairs at Boston. Riflemen proved to be a mixed bag. Camp life was dull, forays and skirmishes with the British were infrequent and even less consequential.[22] With their highly touted weapons practically neutralized by siege tactics, some enlisted personnel among the riflemen grew bored and sullen. Their tempestuous dispositions vexed Washington and his staff. The admiration of Washington and others for riflemen in August gave way to criticism in October, because raucous and unlettered riflemen ignored military protocol and their pretentiousness caused resentment. After a mob of Pennsylvanians broke a companion out of the Prospect Hill guardhouse, observers gave vent to their disgust. One New Englander characterized riflemen as "mutinous" and "vicious"; General Charles Lee "damned them and wished them all in Boston," while Washington "said he wished they had never come."[23] Benjamin Thompson, later Count Rumford, scoffed that "of all useless sets of men that ever encumbered an army, surely the boasted riflemen . . . are most so."[24] Indeed, riflemen had done little to warrant the confidence many, especially the disappointed Washington, had in them.

Yet Washington realized, if somewhat belatedly, that Boston was not the best place to test the rifle's effectiveness. Light infantry tactics based on mobility and selective, accurate fire could not be used there. Although disgusted with the behavior of some enlisted men in the rifle companies, he kept his faith in their weapon. Accordingly, on January 1, 1776, the Pennsylvania rifle regiment was redesignated the First Regiment of the Army of the United Colonies, reflecting Washington's desire to mold his army around the riflemen. That same month Congress directed Pennsylvania to raise five new Continental regiments and specified that each regiment have one company of riflemen.[25] Both Washington and Congress wanted riflemen to comprise a significant portion of the "national" army. Apparently they still hoped to capitalize on the rifle's inherent superiorities.

Congress recognized that sending more than twelve hundred riflemen to Boston had just skimmed the surface of a vast reservoir. Riflemen appeared in abundance in the Continental, state, and militia levies assembled in 1775 and 1776 from New York to Georgia. Pennsylvania, for example, raised an additional one thousand riflemen for state service in March 1776.[26] In short, there were many more rifles available, and many more riflemen under arms, than is normally appreciated.

Greater numbers, however, did not produce greater results. The riflemen at Boston compiled a mediocre record. Three of those rifle companies later went on the Quebec expedition. Instead of filling their enemy with dread, most of the riflemen were taken prisoner during the New Year's Day assault on the city. Because they fought along narrow streets in a drizzling downpour, they held no noticeable advantage over their opponents. On the contrary, faster loading and less temperamental muskets equipped with bayonets were better for house-to-house fighting. The edge riflemen might have enjoyed if the battle had been fought in the open was negated once they passed within the city walls. Rather than exploiting their weapon's superiorities, they became victims of its inadequacies. Rifles had to be employed more imaginatively to be effective, otherwise all of the patriots would have been better off with muskets.

Continental riflemen in South Carolina compiled a slightly better record than their counterparts farther north. When Sir Henry Clinton made his bid to take Charleston in June 1776, South Carolina riflemen acquitted themselves well. Indeed, Charles Lee, commanding at Charleston, showed that the source of his irritation in Boston had been riflemen, not their rifles. He counted heavily on rifle regiments because "the enemy entertain a most fortunate apprehension of American riflemen."[27] They did not have much of a chance to prove their mettle, although they did prevent an amphibious assault from turning the flank of Fort Sullivan.[28]

Interestingly enough, militia in Virginia and North Carolina made the first significant use of the rifle. In December 1775 Virginia riflemen turned out with other Virginia troops to maul a combined British and loyalist force at Great Bridge. Three months later North Carolina riflemen trounced a column of backcountry loyalists at the battle of Moore's Creek Bridge (and took fifteen hundred rifles as booty).[29] At both engagements the patriots picked the site of battle, fought from concealment, and left themselves an easy line of retreat. Because their opponents had to approach along a narrow front, their accurate rifle-fire held sway. The Virginians and North Carolinians maximized the rifle's advantages and avoided a situation where its lack of a bayonet and slowness of loading would be factors.

Until the raising of a special corps under Daniel Morgan in 1777, Continental riflemen rarely had such opportunities. And while Morgan's contingent is often pointed to as the high-water mark for riflemen during the war, rifles were already being phased out of the Continental army several months earlier. There were never more riflemen in the Continental army than in the summer of 1776. The New York campaign of that season would be the last time riflemen comprised a significant portion of the army. Washington had perhaps two thousand riflemen at New York. Present were some New York militia riflemen, the original Pennsylvania rifle regiment (less two companies taken at Quebec), a new though incomplete regiment of Virginians and Marylanders (also minus a company lost at Quebec), two Pennsylvania state rifle regiments, plus the rifle companies in each Pennsylvania and Virginia line regiment. Washington, having forgiven the sins of the riflemen at Boston, would have liked to have had more. At his request Congress induced the original Pennsylvania riflemen to reenlist for a bounty.[30] Washington and Congress, then, still planned to keep riflemen as a significant part of the Continental army. Considering the way those riflemen were used, one might wonder why.

American commanders on Long Island did not use riflemen effectively. Riflemen accounted for fully one-third of the twenty-eight hundred front-line troops stationed there in August 1776, but they were split into small groups. When General William Howe unexpectedly seized the initiative and a British and Hessian column outflanked the American defenses, riflemen had no chance to capitalize on their superior range.[31] One participant in the confused battle that followed noted that German *jaegers*, like American riflemen, did not have bayonets, but unlike Americans, *jaegers* were skilled in linear as well as irregular tactics.[32] Psychologically unprepared to deal with bayonet-wielding regulars, numerous riflemen simply broke and ran. German mercenaries therefore dismissed them as contemptible. "Riflemen were mostly spitted to the trees with bayonets," jeered one Hessian officer, adding that "these frightful people deserve more pity than fear."[33] A British officer later noted that his comrades went out "rebel hunting" at night. Before unlucky riflemen could reload after their first shot, they were "run through . . . as a rifleman is not entitled to any quarter."[34] Whatever mystique had been formerly associated with their prowess had rapidly worn off.

Admittedly, riflemen fared somewhat better on Manhattan Island than they had on Long Island. They won a few minor skirmishes there.[35] Skirmishes did not alter the course of the campaign, however. Washington abandoned Manhattan, moved up to White Plains, crossed into New Jersey, and ultimately retreated into Pennsylvania. The Maryland-Virginia rifle regiment did not join the retreat. It fell captive to the British, along with the rest of the garrison left isolated at Fort Washington by the main army's withdrawal. The dogged resistance of those riflemen turned out to be one of the few bright spots during the entire siege. They fought in open order in hilly terrain north of the fort, sniping at their attackers and inflicting frightening casualties. But because the Hudson was on their flank, they could not withdraw, and they eventually laid down their arms.[36] Tactically they fought as was their wont; strategically, the British dictated the terms of battle.

Washington ended 1776 with the tattered remnants of an army. Victory at Trenton and a fortunate escape through Princeton left him mulling over his prospects for the coming year. Of the two thousand or so riflemen in the army six months earlier, fewer than a quarter remained. A growing number of officers agitated for their elimination altogether. Peter Muhlenberg, colonel of a Pennsylvania line regiment, requested that the men in his command be uniformly armed with muskets.[37] Anthony Wayne complained, "I don't like rifles—I would rather face an Enemy with a good Musket and Bayonet without ammunition." If Wayne had his way, he would see "Rifles Intirely laid aside."[38] The like-minded Board of War hesitated to accept any new rifle companies. If enough muskets had been available, it "would speedily reduce the number of rifles" and replace them with muskets, "as they are more easily kept in order, can be fired oftener, and have the advantage of bayonets." Washington joined the Board of War and his subordinate officers in favoring a substitution of muskets for rifles in line regiments.[39]

Riflemen in a sense became scapegoats for general defects in the American army. Most American soldiers, not just riflemen, could not match British regulars. For both military and political reasons, the Continental army had not fought a truly "revolutionary" war, and the patriots employed weapons tactics that underlined the need for iron discipline, rigorous training, and standardized equipment. Whether or not Washington and Congress had ever seriously intended to fight a shock action, guerrilla war is debatable. If not, then their reliance on riflemen seems to have been based on the misconception that bayonetless rifles and backwoodsmen unfamiliar with linear tactics could win battles where those tactics were employed.

As time wore on, the American army became more like that of its British foe in form and function. In the winter of 1777-1778, esteem for the rifle reached its nadir. That winter spent at Valley Forge witnessed significant changes in the American army. Rifles had steadily fallen in reputation since the halcyon days of 1775, but wholesale disavowal of those weapons came only with the teachings of Baron Friedrich Wilhelm Augustus von Steuben. Von Steuben sought to professionalize an amateurish army. The potpourri of weapons in American regiments, from muskets to rifles to fowling pieces, dismayed him. He corrected that by eliminating everything except bayonet-equipped muskets, insofar as stocks on hand would allow. He taught the manual of arms, platoon volley fire, and proper use of the bayonet.[40]

The "new" American army emerging in the spring of 1778 was an army built around von Steuben's staid European principles. Von Steuben largely deserves credit for transforming "rag, tag and bobtail" into a cohesive, disciplined fighting force. Yet his improvements entailed a rejection of most facets of native American warfare, a shift to the "tried and true" fundamentals of eighteenth-century European warfare. Von Steuben merely formalized a tendency already present. Prior to 1778 American military leaders had not implemented a systematic approach to war. A smattering of European textbook tactics had been meshed with dicta of frontier warfare to produce soldiers comfortable with neither. Inclusion of riflemen in the Continental army in 1775 and 1776 reflected Washington's desire to integrate the best aspects of frontier warfare into his battlefield tactics. That integration failed. Washington had men adept at linear tactics or irregular tactics, but few were proficient at both. Riflemen came least prepared to face the British, not because they did not know how to fight, but because they only knew how to fight in one particular fashion. Those most inclined to use rifles were also the least inclined to fight British regulars on the latter's terms. When faced with a crumbling army, American leaders understandably fell back on European techniques instead of experimenting with something new. In 1778 Charles Lee proposed an alternative to the von Steuben plan, but by then the trend could not be reversed.[41]

A new approach would have required a more eclectic borrowing from European and American military experience. Back in 1757 New England pastor Gad Hitchcock proposed just such a mixture. Hitchcock stressed that the well-trained colonial be adept at both European and Indian warfare. He should "not be unacquainted with the Methods of War that are practised by the Enemy"; he should be able to "fight skillfully, either in the Wilderness or the field."[42] Though addressing himself to prospective militiamen in the French and Indian War, Hitchcock might have offered the same advice in 1775. Hitchcock sought to put the colonial soldier on a par with his opponent, be he Indian or European. The rifle would have given an added dimension. A bayonet-equipped rifle would have held the edge, its user fighting at long-range whenever possible but being able to engage the British at close quarters if necessary.

Daniel Morgan's special rifle corps was the closest the Continental army came to filling Hitchcock's prescription. In June 1777 Washington authorized Morgan to assemble a light infantry regiment of five hundred riflemen.[43] The hulking Virginian scoured the ranks and selected men primarily from Pennsylvania and his home state. (Considering the difficulty he had in finding rifles, the new regiment probably stripped the main army of them.) Morgan's rifle corps was treated as an elite body, as indeed it was. Some of the men in the new regiment had marched with the original rifle companies to Boston in 1775, and they had since been seasoned by experience. Though none had modified their guns to take bayonets, they were not as prone to panic. Henry Knox valued the men in the regiment as the "most respectable body of Continental troops that were ever in America."[44]

In August Washington detached the rifle corps and sent it to assist the northern army under Horatio Gates. Washington informed Governor Clinton of New York that Morgan's men were the "pick of the army." He asked Israel Putnam to exaggerate the number of men with Morgan, hoping the Indians serving with John Burgoyne, on his way down the Hudson, would lose heart and go home.[45] At the battles of Freeman's Farm on September 19 and Bemis Heights on October 7, when Burgoyne tried desperately to turn Gates's flank, riflemen distinguished themselves again and again. Morgan and his regiment fought from concealment, letting loose a withering fire in both engagements. Burgoyne later confessed in testimony before the House of Commons that the riflemen slew an inordinate number of his officers and caused dozens of Indians and loyalists to desert him.[46] William Digby of the Shropshire Regiment observed that at Freeman's Farm all but one of the officers in his regiment fell to the riflemen.[47] At Bemis Heights a British sergeant lamented that "the riflemen from trees effected the death of numbers," including the talented and popular General Simon Fraser.[48]

Morgan's men bested light infantry, grenadiers, and *jaegers*, the cream of Burgoyne's army. Yet they did not fight alone. They had been reinforced by veteran units armed with muskets and bayonets. If not for their support, the riflemen would have been driven from the field at least once during the fighting at Freeman's Farm. Morgan's men may have been the catalyst bringing success in the Saratoga campaign, but they did not win it single-handedly. They constituted a small portion of Gates's eleven-thousand-man army. And despite the lessons of two years of war, their guns could not take bayonets. Either riflemen refused to modify their weapons, fearing they would have to fight in close order, or Gates and his staff did not see how rifles could have been made more satisfactory weapons.

Morgan returned to the main army as a conquering hero. But like the army in general, his corps was decimated by the severe winter of 1777-1778 and some sharp actions with the British. Finally, in July 1778, Washington sent a portion of the regiment west for frontier duty and disbanded the remainder, ordering the men to return to their old units. This regimental reorganization not only ended the chapter on Morgan's contingent, it for all intents and purposes closed the book on the rifle in the Continental army.[49] The irony of this was that it followed less than a year after Morgan's Saratoga triumph.

The days of Morgan's rifle regiment had been numbered from the beginning. A special light infantry corps equipped with muskets and bayonets had been organized in July 1778, at the very time Washington

12. Daniel Morgan, the rifleman as war hero. Courtesy of the Library of Congress.

ordered the fragmenting of Morgan's unit. It would have been assembled earlier if trained men and adequate supplies had been available. Washington ordered each line battalion to organize a light infantry company, the individual companies in each battalion to combine as an independent regiment during campaigns.[50] In other words, Washington essentially reinstituted the system he had pushed for in 1775, except that he replaced riflemen with troops armed and trained to fight in the same manner as regular line troops. Considering the tactics of the Continental army by 1778, that seemed the logical decision. Washington overcame the handicap of having two types of soldiers—riflemen and musket-equipped regulars—in the same army. It was his formal announcement that the Continental army was more European than American.

The virtual disappearance of the rifle from the Continental army did not mean that rifles were no longer used in the war. In the South, Continental riflemen were being phased out in favor of musket-armed regulars by 1777, as in the main army under Washington, but many of those men ended up in the partisan corps of Andrew Pickens, Thomas Sumter, and Francis Marion. Southern militia in fact made the best use of the rifle during the war. Battle lines and full-scale engagements were few and far between, and the rifle finally came into its own. Riflemen won a few striking victories. A small company of Virginians stunned Banastre Tarleton at Wahab plantation; another band smashed a loyalist force three times its size at Musgrove's Mills.[51] Riflemen in the Deep South participated in every action from Fort Watson to Ninety-Six to the last pitched battles of the war.[52] Militia riflemen, many of them veterans of the defunct Continental rifle regiments, turned out to fight at Cowpens and Guilford Courthouse and later marched alongside Lafayette in Virginia.[53]

Riflemen fought most successfully at King's Mountain in October 1780 and Hannah's Cowpens in January 1781. At the former, approximately eleven hundred Tennesseans, North Carolinians, and Virginians, almost all of whom carried rifles, crushed a slightly smaller and similarly armed loyalist army in the largest single action of the war between two bodies of riflemen. The loyalists, many of whom had modified their weapons to take bayonets, tried to decide the battle with a headlong charge that the patriots parried by dispersing and fighting from behind rocks and trees.[54] Giving way before the loyalist onslaught, the backwoodsmen cut their opponents to pieces. They won convincingly with a well-directed fire.[55]

At the Cowpens, Daniel Morgan successfully mixed rifle with line tactics. His Virginia, North Carolina, and Georgia riflemen poured a galling preliminary fire into the British before withdrawing to the rear. Nonetheless, Morgan's victory over Banastre Tarleton resulted as much from good fortune as sound tactics. If not for a sudden wheeling movement and bayonet charge by the Continentals, the steadfastness of the Virginia riflemen, the unexpected return to the field of the other militia, and a slashing cavalry charge led by William Washington, the scales could have tipped to the other side.[56] Morgan's fortuitous mixture of skittish militia and dependable line troops made his gamble pay off. Yet if he had had an army capable of fighting in either irregular or line fashion, depending on the situation at hand, his battle plan would not have been so risky. Nathanael Greene's unsuccessful attempt at Guilford Courthouse to imitate Morgan demonstrated just how lucky the latter had been.

Greene's problem at Guilford Courthouse reflected the idiosyncrasies of the American military establishment. Militia and Continentals waged two different types of war. This explains how Washington could phase the rifle out of the Continental army on the one hand and on the other advise New York to raise a regiment of militia riflemen to serve on the frontier.[57] Militia, particularly when called to fight outside their home state boundaries, had a disturbing habit of coming and going as they pleased. Still, because of their predilection for hit-and-run tactics, militia fought in a way making the rifle useful to them. Continentals had little use for temperamental, bayonetless rifles. They dressed, drilled, and fought much like their British foes. They may in fact have become too much like them. One American officer wrote that at an encounter near Green Springs, Virginia, just before Cornwallis pinned himself down in Yorktown, some British troops ironically turned the tables on their American attackers. American light infantry, bayonets leveled, advanced in close order through a woods only to be stopped and hurled back in confusion by British regulars firing individually while dodging from tree to tree.[58]

The natural disparity between militiaman and Continental was institutionalized by the American military organization. Continentals and militia were recruited and brigaded by state, with few exceptions. Early attempts to replace this system with a truly national army went nowhere. Getting all elements of such a diffused organization to fight in a new way would have been a herculean task.

Military conservatism also played a role in curtailing the rifle's use. To be sure, anticipating and planning for war before April 1775 was not feasible, given the evolutionary nature of agitation for a break with the mother country. That the patriots had to learn from experience is to be expected. Somewhere along the line, however, some farsighted strategist might have seen that the rifle was perhaps too perfect a reflection of colonial frontier warfare. It needed to be modified to serve against an enemy of a very different nature or else laid aside as unsuitable. Modification entailed changing the backwoodsman's aversion to training in bayonet tactics as well as physically altering the rifle. After all, the war would not have been won any sooner if the entire army had been composed solely of frontier riflemen lacking that training.[59] John Simcoe, commander of the Queen's Rangers, correctly charged that riflemen, because of their limited training, "were by no means the most formidable of the rebel forces."[60]

Washington and his staff did not have a free hand at making strategy. Their troops could not simply melt away into the woods with the hope that the British would follow, there to be shot to pieces by hidden marksmen. Washington had political considerations he could not ignore. He had to keep his army close to the seaboard and, as at New York, sometimes to stand and fight when his instincts told him to withdraw. Battle sites better suited to the rifle may not have worked either. There is the distinct possibility that irregular tactics based on the rifle may have led the patriots to avoid a serious confrontation with the British, thereby reducing the Continental army's effectiveness. That army had to be an active deterrent to British designs, if only to keep enlistments from dropping. Americans could not have fought a partisan, guerrilla action. Indeed, William Moultrie wrote after the war that Fabian tactics caused too many to "grow tired and desert."[61] American soldiers were poorly disciplined and often wandered off for home; irregular tactics may have made a bad situation worse.[62]

Most American leaders eventually agreed that the rifle needed to be improved, but they did not attempt to introduce any changes themselves. Practically no inventive interest was taken in the rifle during the war. David Rittenhouse and Charles Willson Peale experimented with a telescopic sight for rifles in 1776 and ended their work abruptly when they almost put out their eyes. Rittenhouse later proposed to experiment with rifled cannon, but nothing came of his proposal.[63] The patriots were not equipped for sustained wartime experimentation.

American leaders ultimately gave up on the rifle. By war's end they had almost nothing good to say about it. Some did not start out that way, otherwise Washington would not have attempted to fill the army with riflemen in 1775 and early 1776. At first glance Washington's later change of heart could be viewed as a contradiction to his professed faith in the rifle. He, along with countless others, had made the rifle a national symbol, hard evidence that the colonies had bested the mother country in one field of technological endeavor. Nonetheless, champions of that weapon found themselves at the edge of the same void as those who championed pre-war home manufacturing when converting their thoughts to action. Their zeal outstripped their ability; faith alone was no substitute, and that faith ultimately faltered. The patriots expected great things from the rifle, but when those great things did not materialize, they retreated to an imitative, Europeanized approach to war. When riflemen failed to produce the desired results, few understood why. Adapting the rifle to tactical realities posed one set of problems. The intellectual gymnastics of rethinking tactics to maximize the rifle's effectiveness posed another, whose solution lay beyond the ken of the Revolutionary generation.

Thus the Pennsylvania rifle, a successful adaptation to one environment, did not fare as well when placed in another. It was more than a tool. It reflected a certain attitude about war, an attitude not universally applicable. American military leaders did not successfully adapt the rifle to their tactics or their tactics to the rifle. They had had limited options. At no point during the war did they have the luxury of retooling, because Congress and the states operated with marginal resources. Benjamin Franklin's atavistic proposal that pikes, bows, and arrows replace firearms as standard weapons resulted as much from the constant munitions shortages as from Franklin's dislike of muskets.[64] Even if Washington and Congress had committed themselves to a new kind of army equipped with rifles, they probably could not have carried if off. Pennsylvania gunsmiths would have been happy to try, since they welcomed "an excuse to lay by the Musketwork and make rifles, which are more profitable for them," but rifles took more time to make than muskets, and undoubtedly many gunsmiths assembling muskets under committee of safety and congressional contracts were unfamiliar with the art.[65] Besides, it proved impossible to keep Americans adequately supplied with muskets much less rifles. Not only were rifles costlier and harder to make, European gunsmiths producing a large percentage of the patriots' arms were not acquainted with rifles, or at least with American models.

Whether the rifle would have been used more imaginatively if the patriots had had the industrial capacity to produce nine or ten thousand a year is a moot point. Though neither Washington nor Congress said so explicitly, the realization that they had to fight with whatever they could scrape together on short notice may have shaped their thinking on the rifle. Be that as it may, it can be safely stated that the Pennsylvania rifle, though the product of American genius, was not ingeniously employed during the War of Independence. What is more, given the state of American industry and the tendencies of American troops, it is just as well that the rifle experiment was set aside.

The rifle illustrates the limits of successful technological innovation during the war. Americans did not win the war with the rifle. They had attempted to, however, which is even more significant. By turning to the rifle, they demonstrated their willingness to innovate. They realized that technology could be a source of power. Inability to capitalize on the rifle, like inability to produce adequate munitions stocks, brought renewed determination for technological change. That determination replaced the disappointment that engendered it. Advocates of technological change after the war became more insistent than ever. They could point to accomplishments like those of the Delaware engineers to show that they had the necessary genius, and they could point to the rifle to show that they could innovate.

Just as independence gave Americans the opportunity to experiment with new political theories and forms, it gave them the opportunity to experiment technologically. That experimentation proceeded after the war in dead earnest because the Franklins, the Washingtons, and the Hamiltons in American society knew that their nationalistic ambitions far outstripped their technological aptitude. If they did not want to surrender the former, they had to improve the latter.

NOTES

1. Burlingame, *March of the Iron Men*, p. 121; echoed in John A. Kouwenhoven, *Made in America* (New York: Doubleday and Co., 1948), pp. 14-15. Like statements are in Heller, "Gunmakers of Old Northampton," p. 6; and Harry P. Davis, *A Forgotten Heritage* (Huntington, W.V.: Standard Pub. Co., 1941), pp. 82-93. At the prompting of George Shumway, author of *Rifles of Colonial America*, 2 vols. (York, Pa.: George Shumway, 1980), I have revised some of my earlier statements about the evolution of the rifle in the colonies. Given the spread of rifles to areas outside of Pennsylvania, Shumway suggests that "American longrifle" is more accurate and descriptive than "Pennsylvania rifle." Moreover, Shumway and other recent writers have noted that European rifles underwent many changes, just as the American version evolved after 1700. What distinguishes European from American types is response to practical need, not unimaginative European versus ingenious American.

2. Burlingame, *March of the Iron Men*, p. 133.

3. Robert Held, *The Age of Firearms* (New York: Harper and Row, 1957), p. 138. In 1635 the first patent for rifling was granted in England; see James E. Severn, "The Rifled Bore, Its Development and Early Employment," *The American Rifleman* 110 (March 1962):30. Benjamin Robins was probably the first Englishman to make extensive experiments with rifled pieces, however. For his report to the Royal Society of London of 2 July 1747, in which he predicted the rifle would revolutionize warfare, see James Wilson, ed., *The Mathematical Tracts of the Late Benjamin Robins*, 2 vols. (London: J. Nourse, 1761), I, 328-341.

4. Nonetheless, greased patches were not widely used in Europe until later because many subscribed to the "retarding and resisting" theory of rifling, which postulated that improved velocity and range came from the friction and compression generated when the ball was mashed down by a mallet. See Held, *Age of Firearms*, p. 139; and Henry J. Kauffman, *The Pennsylvania-Kentucky Rifle* (Harrisburg: Stackpole Co., 1960), p. 2.

5. See George Shumway, *Rifles of Colonial America*, passim; Felix Reichmann, "The Pennsylvania Rifle: A Social Interpretation of Changing Military Techniques," *PMHB* 69 (1945):3-14; Carlton O. Wittlinger, "The Small Arms Industry of Lancaster County, 1710-1840," *Pennsylvania History* 24 (January-October 1957):121-136; John C. Dillin, *The Kentucky Rifle* (Washington, D.C.: National Rifle Association, 1924); Horace Kephart, "The Rifle in Colonial Times" *Magazine of*

History 24 (1890):79-81; and Kauffman, *Pennsylvania-Kentucky Rifle*, pp. 8-31. See Francis Jordan, *The Life of William Henry* (Lancaster: Press of the New Era Printing Co., 1910), pp. 3-55, passim, for one of the most important gunsmiths.

6. Charles W. Sawyer, *Firearms in American History* (Boston: Charles Sawyer, 1910), pp. 153-157; and Townsend Whelen, *The American Rifle* (New York: Century Press, 1918), p. 6; also see Peterson, *Arms and Armor in Colonial America*, p. 155.

7. J. F. C. Fuller, *British Light Infantry in the Eighteenth Century* (London: Hutchinson and Co., 1925), pp. 46-70, passim.

8. Eric Robson, "The Armed Forces and the Art of War," *The New Cambridge Modern History* 7 (1957):174.

9. Ballagh, ed., *Letters of Richard Henry Lee*, I, 130-131.

10. Burnett, ed., *Letters*, I, 134. Also see the letter from John Adams to Elbridge Gerry of 18 June 1775 in Smith, ed., *Letters of Delegates* I, 503.

11. See John William Ward, *Andrew Jackson: Symbol for an Age* (Oxford: Oxford University Press, 1953), pp. 13-29, for a discussion of how Americans of the next century expanded the heroic image of the riflemen.

12. Ford, ed., *JCC*, II, 89.

13. Ibid., II, 104, 173; also see Smith, ed., *Letters of Delegates*, I, 491.

14. Aaron Wright, "The Journal of Aaron Wright," *Boston Evening Transcript*, 11 April 1862. Thompson's itemized expenses are in Force, ed., *Amer. Archives*, 4th series, III, 1045-1046; Morgan's are in Ford, ed., *JCC*, III, 267, 319-320, 329, 370; and other company commanders are elsewhere in ibid. A note on the recruitment of the Maryland companies is in the *Pennsylvania Packet*, 10 July 1775.

15. *Va. Gazette* (Purdie), 19 August 1775.

16. *Boston Gazette*, 14 August 1775.

17. George Morison, "Journal of the Expedition to Quebec," in Kenneth Roberts, ed., *March to Quebec* (Garden City: Doubleday and Co., 1938), pp. 506-508. Flattering pieces on the riflemen include Horace Kephart, "The Birth of the American Army," *Harper's Monthly Magazine* 98 (1899):961-970; and William W. Edwards, "Morgan and His Riflemen," *WMQ*, 2nd series, 23 (1914):73-106. Also see Henry J. Young, "The Spirit of 1775," *John and Mary's Journal*, no. 1 (March 1975); and Joe D. Huddleston, *Colonial Riflemen in the American Revolution* (York, Pa.: George Shumway Publisher, 1978), for the riflemen from Boston to Yorktown, including those in the West with George Rogers Clark. The *Pennsylvania Packet*, 28 August 1775, postscript, quoted a source that put the total number of riflemen at 1,430. Like many other papers, the *Packet* followed the exploits of the riflemen throughout the summer of 1775.

18. John Joseph Henry, *Account of Arnold's Campaign Against Quebec* (Albany: Joel Munsell, 1877), p. 11; Thacher, *Military Journal*, p. 31; and James Graham, *Life of Daniel Morgan* (New York: Derby and Jackson, 1858), p. 63.

19. Fitzpatrick, ed., *Writings of Washington*, III, 325, 404, 415; V, 336; and II, 229, for Washington's feelings back in 1758.

20. Thacher, *Military Journal*, p. 31; an earlier demonstration by a Virginia company is noted in Force, ed., *Amer. Archives*, 4th series, III, 2; and John Harrower, "Diary of John Harrower, 1773-1776," *American Historical Review* 6 (1900):100.

21. *New York Journal*, 17 August 1775; Caleb Haskell, "Diary at the Siege of

Boston and on the March to Quebec," in Roberts, ed., *March to Quebec*, pp. 467, 468-472; Fitzpatrick, ed., *Writings of Washington*, III, 393-394; IV, 84; Heath, *Memoirs*, p. 18; and Frank Moore, ed., *Diary of the American Revolution* (Hartford: J. B. Burr Pub. Co., 1876), pp. 119-120.

22. Edward Hand, second in command of the Pennsylvania rifle regiment, admitted this. See his letter of 29 August 1775 in the Edward Hand Papers, Force Transcripts, Series 7E, I, p. 5, LC MSS.

23. For the Prospect Hill incident, see Commager and Morris, eds., *Spirit of Seventy-Six*, pp. 156-157; and Fitzpatrick, ed., *Writings of Washington*, III, 490-491. For derisive comments about the riflemen, see the Adams Papers, Artemas Ward to John Adams, 30 October 1775, I, 98 (reel 345); John Thomas to John Adams, 24 October 1775, I, 101 (reel 345); and Samuel Osgood to Adams, dated 25 October 1775, I, 103 (reel 345). These letters have now been published in vol. III of Robert J. Taylor et al., eds., *Papers of John Adams* (Cambridge: Harvard University Press, 1979), pp. 231-241. Adding substance to the charges, see the list of deserters from among the riflemen in the GWP, series 4, entry of 23 October 1775 (reel 34). And it would seem that not all of the riflemen were truly marksmen, because some were weeded out and sent home. See Fitzpatrick, ed., *Writings of Washington*, III, 490; and muster rolls in Force, ed., *Amer. Archives*, 4th series, III, 253-254; IV, 491-492.

24. Commager and Morris, eds., *Spirit of Seventy-Six*, p. 155.

25. Ford, ed., *JCC*, IV, 29. The designation of the original rifle regiment was changed to the First Regiment of the Pennsylvania Line in July 1776, when a state numbering system was adopted.

26. Force, ed., *Amer. Archives*, 4th series, V, 677, 681, 1225; William H. Egle, ed., *Pennsylvania in the War of the Revolution*, 2 vols. (Harrisburg: E. K. Meyers, 1890), I, 263; and Labaree, ed., *Papers of Franklin*, XXII, 189, for riflemen in the York County, Pennsylvania, militia. For the creation of a Maryland-Virginia regiment of Continental riflemen, see Fitzpatrick, ed., *Writings of Washington*, V, 202; Ford, ed., *JCC*, V, 433, 542, 486; and the individual companies in Force, ed., *Amer. Archives*, 5th series, I, 92-93, 95, 131-132, 183.

27. Jared Sparks, ed., *Correspondence of the American Revolution*, 4 vols. (Freeport: Books for Libraries Press, 1970), II, 501-502.

28. Edward McCrady, *The History of South Carolina in the Revolution, 1775-1783*, 3 vols. (New York: Paladin Press, 1969), III, 152-153.

29. Great Bridge in Force, ed., *Amer. Archives*, 4th series, IV, 501-502; Gordon, *History*, II, 111-113; Moore's Creek Bridge in ibid., II, 209; the *Pennsylvania Magazine* 2 (April 1776):196; and Ward, *War of the Revolution*, II, 663-664.

30. Fitzpatrick, ed., *Writings of Washington*, IV, 501-502. The original Pennsylvania rifle regiment (now the First Regiment of the Pennsylvania Line) was severely under strength, however—see Force, ed., *Amer. Archives*, 5th series, I, 331-332. See Ford, ed., *JCC*, V, 473, for attempts to raise new companies. See Fitzpatrick, ed., *Writings of Washington*, V, 202, 216; and Burnett, ed., *Letters*, I, 518, for the under-strength status of the Virginia-Maryland regiment.

31. See Henry Johnston, *The Campaign of 1776 Around New York and Brooklyn* (Brooklyn: Long Island Historical Society, 1878), pp. 154, 163, 166, 182-185; and Part 2 (Documents), pp. 64-65; also Force, ed., *Amer. Archives*, 5th series, I, 1213-1214; Commager and Morris, eds., *Spirit of Seventy-Six*, pp. 433-440.

32. Johnston, *Campaigns*, Part 2 (Documents), p. 50.

33. Edward J. Lowell, *The Hessians* (Port Washington: Kennikat Press, 1965), pp. 65-66.

34. Moore, ed., *Diary of the Amer. Rev.*, pp. 349-350.

35. Rifleman skirmishes at Harlem Heights are noted in Commager and Morris, eds., *Spirit of Seventy-Six*, pp. 470-471; Johnston, *Campaigns*, Part 2 (Documents), pp. 86-87; idem, *The Battle of Harlem Heights* (New York: AMS Press, 1970), pp. 54-55; Fitzpatrick, ed., *Writings of Washington*, VI, 146, 179; Throg's Neck in Ward, *War of the Revolution*, I, 255; and Heath, *Memoirs*, pp. 59-60; skirmish at Mamaroneck in Lowell, *The Hessians*, pp. 75-76.

36. Alexander Graydon, *Memoirs of His Own Times* (New York: New York Times and Arno Press, 1969), pp. 192-202; John W. Fortescue, *A History of the British Army*, 13 vols. (London: Macmillan Co., 1899-1920), III, 191-193.

37. Letter from Muhlenberg to Washington of 23 February 1777 in the GWP, series 4 (reel 40). Also see John W. Wright, "The Rifle in the American Revolution," *American Historical Review* 29 (1924): 293-299.

38. Wayne to the Board of War, 3 June 1777, in the General Anthony Wayne Papers, III, p. 89, HSP MSS.

39. Force, ed., *Amer. Archives*, 5th series, II, 1247. The Washington comment can be found in a Wayne letter to Richard Peters of 8 February 1778 in the Wayne Papers, IV, 78, HSP MSS. Also see the letter from Colonel William Thompson of South Carolina requesting that most of his rifles be replaced with muskets in Burnett, ed., *Letters*, II, 452.

40. John M. Palmer, *General von Steuben* (New Haven: Yale University Press, 1937), pp. 140, 151-157. The enthusiastic reception of von Steuben's modifications is noted in a letter from Timothy Pickering, a member of the Board of War and himself author of a military handbook, to Congress of 27 March 1779 in the PCC, Item 147, III, 143 (reel 158).

41. Lee in the Lee Papers, vol. 2, *New York Historical Society Collections. John Watts De Peyster Publication Fund* 5 (1892):383-389.

42. Gad Hitchcock, *Sermon* (Boston: Edes and Gill, 1757), p. 12. For the conventional, non-revolutionary character of the war, see John Shy's essays, "American Strategy: Charles Lee and the Radical Alternative" and "The Military Conflict Considered as a Revolutionary War," reprinted in his *A People Numerous and Armed* (New York: Oxford University Press, 1976), pp. 133-162, 193-224.

43. Fitzpatrick, ed., *Writings of Washington*, VIII, 156, 236-237.

44. Commager and Morris, eds., *Spirit of Seventy-Six*, p. 537.

45. Fitzpatrick, ed., *Writings of Washington*, IX, 70-71, 78, 82, 102.

46. John Burgoyne, *A State of the Expedition From Canada* (London: J. Almon, 1780), pp. 30, 102, 121-122. For commentaries on Morgan and his unit, see Don Higginbotham, *Daniel Morgan* (Chapel Hill: University of North Carolina Press, 1961), pp. 16-77; and John S. Pancake, *1777: The Year of the Hangman* (University: University of Alabama Press, 1977), pp. 82-83, 146-187.

47. Commager and Morris, eds., *Spirit of Seventy-Six*, p. 580; also see James Wilkinson, *Memoirs of My Own Times* 2 vols. (Philadelphia: Abraham Small, 1816), I, 243-247.

48. Roger Lamb, *Memoirs of His Own Life* (Dublin: J. Jones, 1811), p. 199; also Thacher, *Military Journal*, pp. 101-102; and Stedman, *History*, I, 336-344.

49. Fitzpatrick, ed., *Writings of Washington*, XII, 140, 200, 214, 284, 406; XIII, 110; XIV, 43; XVII, 85. The corps was not officially disbanded until 7 November 1779, when the riflemen returned from western duty. Those riflemen had served well in Sullivan's Indian campaign. Their rifles were handed over to the commissary of military stores, not to be redistributed except by Washington's personal order. Two companies—about ninety men—served in 1780 and 1781 as special sharpshooters. See ibid., XIX, 252, 379, 479; XX, 187, 402. A new regiment, never brought up to strength, was organized for the Yorktown campaign, once again as a special sharpshooting unit. See ibid., XXII, 257-258, 341, 426-427.

50. John W. Wright, "The Corps of Light Infantry in the Continental Army," *American Historical Review* 31 (1926):454-461.

51. Ward, *War of the Revolution*, II, 738-739.

52. Jac Weller, "Irregular But Effective: Partizan Weapons Tactics in the American Revolution, Southern Theatre," *Military Affairs* 21 (1957): 118-131.

53. For riflemen with Greene at Guilford Courthouse, see George W. Greene, *The Life of Nathanael Greene*, 3 vols. (New York: Books for Libraries Press, 1972), III, 189-202; Henry Lee, *Memoirs of the War in the Southern Department* (Washington, D.C.: Peter Force, 1827), pp. 170-180; and Banastre Tarleton, *A History of the Campaigns of 1780 and 1781 in the Southern Colonies of North America* (London: T. Cadell, 1787), pp. 269-279, 303-312; and with Lafayette in Gaillard Hunt, ed., *Fragments of Revolutionary History* (Brooklyn: The Historical Printing Club, 1892), pp. 29-40, 46-56.

54. According to Stedman, *History*, II, 220-223, many of the loyalists at King's Mountain had modified their rifles to take bayonets. See the general accounts in Commager and Morris, eds., *Spirit of Seventy-Six*, pp. 1135-1145; Lyman C. Draper, *Kings Mountain and Its Heroes* (Chapel Hill: University of North Carolina Press, 1967), particularly p. 237; and J. Watts De Peyster, "The Affair at King's Mountain," *Magazine of American History* 5 (1880):401-424.

55. Moultrie, *Memoirs*, II, 245.

56. Accounts are in Tarleton, *Campaigns*, pp. 214-222; Graham, *Daniel Morgan*, pp. 289-316; Commager and Morris, eds., *Spirit of Seventy-Six*, pp. 901-902; and Stedman, *History*, II, 318-325.

57. Fitzpatrick, ed., *Writings of Washington*, XIV, 188. Edward Hand ended up on frontier duty after leaving the Pennsylvania rifle regiment. While at Fort Pitt, his men were equipped with rifles. See Hand's letter of 12 April 1777 to Benjamin Flowers, Commissary of Military Stores, in the Hand Papers, LC MSS. In May 1778, Washington requested various Pennsylvania counties to supply rifles for his troops headed for western service. See the 18 May 1778 letter of George Bryan of Lancaster County to Henry Laurens in the PCC, Item 69, I, 513 (reel 83).

58. Hunt, ed., *Fragments of Rev. History*, p. 50.

59. For claims of this sort, see Sawyer, *Firearms*, pp. 33, 37, 77-79.

60. John G. Simcoe, *Simcoe's Military Journal* (New York: Bartlett and Welford, 1844), p. 237. Likewise, British officer George Hanger, while an admirer of the rifle as a firearm, criticized its limited tactical adaptability. See his *General George Hanger to All Sportsmen* (London: J. J. Stockdale, 1816), pp. 121-124, 199-200. The British were prone to their own brand of military conservatism, passing up a chance to try out in earnest a very effective breechloading rifle invented by Major

Patrick Ferguson. William Howe took advantage of Ferguson's wound at the Brandywine in 1777, where one hundred of the inventor's breechloaders were first revealed, to crate the rifles and store them in a New York City basement.

61. Moultrie, *Memoirs*, I, 365.

62. Daniel J. Boorstin, *The Americans: The Colonial Experience* (New York: Random House, 1958), pp. 368-369.

63. See Maurice Babb, "David Rittenhouse," *PMHB* 56 (1932):113-125; Charles Sellers, *Charles Willson Peale*, 2 vols. (Philadelphia: American Philosophical Society, 1947), I, 126-133; and Force, ed., *Amer. Archives*, 4th series, V, 729. A parallel could easily be drawn between the phasing out of the long bow in the English army and slowness to adopt rifles centuries later. See Thomas Esper, "The Replacement of the Long Bow by Firearms in the English Army," *Technology and Culture* 6 (1965):382-393. Excellent discussions of the general problem of technological development and slow military adaptation are in I. B. Holley, *Ideas and Weapons* (New Haven: Yale University Press, 1953), pp. 3-22; and Elting E. Morison's case study of naval gunnery in his *Men, Machines and Modern Times* (Cambridge: MIT Press, 1966), pp. 17-44.

64. Smyth, ed., *Writings of Franklin*, VI, 438-439.

65. Extracted from Egle, ed., *Penn. in the War of the Revolution*, I, 510.

7 INDUSTRIAL BEGINNINGS

American technology passed through a confused stage during the immediate post-war years. Many Americans seemed to want large-scale technological change more than ever, yet they appeared to be no closer to bringing it about than they had been in 1775. From a short-term economic standpoint, political independence had baleful as well as salutary effects. The new nation freed itself from formal mercantilistic ties with Britain but in doing so cut the umbilical cord of economic sustenance. Now outside the empire, Americans had to adjust to their new freedom. Much like the heart of a coronary victim, they had to depend on alternate routes to keep the economy healthy until they made that adjustment. They were warned by pessimistic forecasters to expect hard times. Europe had not suffered a crippling depression, and the West Indies had not starved, as some patriots predicted they would, when war disrupted the colonial carrying trade. The American colonies had been but a small part of a great world market, an unenviable status inherited by the new nation.[1] As Alexander Hamilton would chide in his 1791 *Report on the Subject of Manufactures*, the United States needed to improve considerably and upgrade an industrial sector that had been a national embarrassment during the war.

Hamilton's report followed hard on the heels of a resurgent home manufactures movement. In some ways the home manufactures craze of the 1780s was identical to the pre-war movement. As before the war, Americans were told to be frugal, to shun ostentatious European finery; and "how to" pieces appeared in magazines, newspapers, and almanacs around the country.[2] Americans were urged to buy homemade wares and to put patriotism first, self-interest second. One magazine listed alphabetically over 220 "raw materials and natural productions" available to the public, trying to show that patriotism would not entail too great a material sacrifice.[3]

Post-war manufacturing enthusiasts also drew on the war itself. If Hamilton did so to criticize, there were others who emphasized the more positive side of wartime production. They played on nationalistic tendencies by honoring the achievements of those craftsmen who had produced goods for the war effort. They reminded the public that during the war, "in the hours of need," craftsmen had produced implements "without which the war could not have been carried on, whereby their oppressed country was greatly assisted and relieved." Any true patriot would therefore tighten his belt and forgo imports, taking a financial loss if necessary to promote the "divers and beneficial arts and manufactures" that helped bring victory.[4] Pre-war emphasis on British oppression was replaced with an emphasis on the need for national accomplishment—a continuation of the war in another form. The military conflict had severed the political tie; a technological contest would bring economic independence.

Orators and civic leaders hammered home to their constituents the need for manufactures. Sir James Jay argued that only manufactures could insure national prosperity; Philadelphia lawyer William Barton echoed Jay and others of the same persuasion.[5] Matthew Carey's monthly magazine, the *American Museum*, quickly became the most vigorous advocate in press for American manufactures. It reprinted articles by Barton, Tench Coxe, and Alexander Hamilton and carried notices of premiums being offered by manufacturing societies. Most issues included statistics on the imports and exports of leading cities, and a section called "Rural concerns" offered tips on scientific farming. Contributors to this and other magazines undoubtedly agreed with "Homespun," who, while conceding industrial prowess could not be achieved overnight, charged that "it is high time we should make a beginning."[6]

And a beginning was made. The post-war movement was far more intense and sustained than the pre-war movement. It eventually brought a more visible return, though that return did not come until the close of the Revolutionary Era. Some Americans capitalized more effectively on certain natural features of their society that stimulated technological change. They better understood that the low proportion of labor to land and westward movement made laborsaving machinery naturally attractive as well as necessary to economic well-being; machines would replace the backs and hands lost to farming. Those Americans also recognized that freedom from imperial rule could open new business opportunities.[7] Manufacturing zealots planned to translate nationwide concern over economic solvency into enthusiasm for technological change. In addition, the transit of technology begun before the war became even more pronounced after it. That transit brought a more heightened interest in mechanization, the very public reaction the manufacturing enthusiasts wanted. Manufacturing enthusiasts therefore sensed that they had a greater chance for success in 1785 than they had had in 1765. "During the connection of this country with Great Britain, we were taught to believe that agriculture and commerce should be the only pursuits of Americans," noted Benjamin Rush, "but experiments and reflection have taught us that our country abounds with resources for manufactures of all kinds."[8] Those resources, Rush believed, needed only to be teamed with new machines to guarantee success.

Of utmost importance, leading Americans took a closer look at technology and recognized that they could not be content with passively borrowing from Europe. Technological borrowing was a good starting point, but no more than that. Americans like Hamilton and even Jefferson, though in a different way, were beginning to see invention as central to technological progress and as an aid in their quest for economic independence. They learned of the role being played by inventors and new devices in fostering the Industrial Revolution in Great Britain. References to Arkwright and Watt, practically non-existent before the war, now turned up in their correspondence.

Not content to wait for British technology to find its way across the Atlantic, American enthusiasts engaged in industrial espionage on a scale unheard of before the war. "Machines appear to be objects of immense consequence to this country," observed the editors of the *Pennsylvania Gazette*. And, they stated bluntly, "It is the duty of every friend to America, at home and abroad, to keep a vigilant eye upon everything of that kind which comes his way." As the editors saw it, "We may invent, and we may borrow of Europe their inventions."[9] It would seem that those newspapermen had a clearer picture of what constituted an important invention than they had had in 1771. They joined others who "had caught the infectious enthusiasm for mechanization."[10]

Philadelphia merchant and political economist Tench Coxe, the postwar manufacturing enthusiast incarnate, warned that the new nation had to develop its own technological expertise as well as adapt European techniques and machines to American conditions. Manufacturers had to press on more aggressively than ever before and adopt water- and steam-powered machinery, he advised. Coxe admitted that initially Americans would have to imitate their former masters, the undisputed leaders in mechanization, but after that, "we may strike out for ourselves."[11] After draining Great Britain of its technological knowledge, Americans should then legislate prohibitive duties, a wall behind which they could improve on that knowledge and march forward to devise technologies of their own. In short, Coxe's ultranationalistic program dictated that the new nation cut loose its technological tie as soon as possible.[12] Until then, manufacturers ought to import people and machines and post premiums to attract inventive minds from abroad. They should subsidize the work of those at home as well.

Coxe lived up to his own challenge, using personal funds in an unsuccessful attempt to bring some Arkwright machinery models into the country.[13] Others endeavored to do the same and smuggle in mechanical devices, since the British, as before the war, were averse to letting outsiders feast off their growing industrial might. Unbeknownst to local authorities, Silas Deane toured English manufacturing centers after the war, where he "examined their late, & new inventions, & machines of various kinds."[14] Deane, like Coxe, did not discover the secrets he sought, nor did other Americans during the 1780s do much better. Nevertheless, their appetite for machine technology had been whetted. As Coxe's partner in industrial espionage, Andrew Mitchell, wrote from London, "The improvements made here upon the different mechanical arts are truly astonishing" and at the root of Britain's economic boom.[15]

The Industrial Revolution tightened its grip on Great Britain in the 1780s. Technology figured prominently in the list of factors bringing economic progress in Britain, as Mitchell and Coxe well knew. Watt's double-acting rotary steam engine, Henry Cort's puddling and rolling of iron, and the addition of Samuel Crompton's "mule" to textile machinery made it "clear that a technological revolution was afoot in Britain."[16] Some American manufacturers and technological enthusiasts like Coxe, wanting the same for themselves, predicted that machine technology would ultimately flourish in a country with a chronic labor shortage produced by a seemingly endless supply of land. Others feared that the population would soon become too great for the towns and countryside to support, and they argued for manufacturing on the basis of labor density rather than on labor scarcity. All ended up competing for British machine technology, most particularly Arkwright-type cotton spinning devices, with a marked intensity.[17]

Coxe recognized that political independence could not have been timed any better, because it came simultaneously with the blossoming of the Industrial Revolution in Britain. In a most fortuitous development, political independence and the desire to achieve economic independence came at the very moment when technological transfer from Britain could help make the dream of industrial might a reality.

Although Tench Coxe waxed most eloquent on manufactures, his program for technological independence also covered agriculture. When he said that the latest machines should be imported, studied, copied, and improved, he meant farm as well as factory implements. He realized that agriculture and manufactures should be pursued as complementary, not conflicting, forms of economic endeavor, both of which had to advance to guarantee the success of the "republican experiment." Even Hamilton admitted that agriculture, "the chief source of subsistence to man," could not be ignored if the nation were to advance industrially.

Metcalf Bowler focused his attention more directly on agriculture, but like Coxe and Hamilton he contended that Americans had to copy and improve on the most progressive European farming techniques.[18] John

13. Tench Coxe, promoter of American technological growth. Courtesy of the Historical Society of Pennsylvania.

Beale Bordley of Maryland, who was to agriculture what Tench Coxe was to manufactures, familiarized himself with the writings of Arthur Young and the Norfolk system practiced by scientific farmers in England. He added insightful observations of his own—to the same end as Bowler, Hamilton, and Coxe.[19]

American agriculturalists were aided in their quest for improvement by European writers. Englishman Charles Varlo compiled a lengthy study of American farming techniques, offering suggestions on better crop strains, planting procedures, and machinery.[20] His hefty tome sat on American library shelves next to Arthur Young's *Annals of Agriculture*, a more polished work by a more polished agronomist. A serialized translation of the Abbé Raynal's writings on agriculture and manufactures caught the eye of progressive American farmers, who also read the numerous letters and scientific papers written by Europeans and reprinted in American magazines.[21] The *American Magazine* noted that Baron Poellnitz, a German scientific farmer of some note, had arrived in New York in the fall of 1788 with "a model of the newly invented [Winlaw's] Threshing mill."[22] Poellnitz prospered at his adopted Long Island home, made the acquaintance of leading Americans, was a key figure in a drive for a national experimental farm, and published a series of essays on farm machinery.[23]

In the long run, émigrés like Poellnitz were perhaps more important to the improvement of American farming than the treatises of those who remained in Europe. And Poellnitz was only one among many talented European farmers to set out for America. Numerous Englishmen joined the host of European farmers who packed up and left. Hundreds of English mechanics wanted to do likewise, but British policy, sometimes ineptly, sometimes effectively enforced, was designed to cut off the flow of manufacturing technology to America. The British government did not block the exportation of agricultural information since, ideally, if Britain maintained its lead in textiles manufactures, more efficient American crop production would feed its factory labor (with corn and wheat) and machines (with flax, hemp, and cotton). Discontented English mechanics, barred from leaving home and seeing the opportunities waiting for them in America, did not accept that logic.

One group of English textile workers determined to leave for America even before the War of Independence had ended. In January 1782, Henry Wyld, a schoolmaster and pattern maker acting on behalf of a party of Manchester textile workers, presented himself to Benjamin Franklin, then in Paris. Wyld and his friends intended to transplant themselves in Pennsylvania and petitioned Franklin for aid. First, they wanted safe-conduct papers for protection against seizure by American or French high-seas cruisers. Second, they asked Franklin to inform the Pennsylvania government of their plans and persuade that body to advance them enough money for the voyage and initial operating costs. Third, they wanted a monopoly on the machines and processes they introduced in the state. Wyld and the others promised to bring "all kinds of Models" of textile machinery and were confident they could duplicate what had been done in Manchester. They also claimed that they had been joined by "some of the greatest Geniusses of our Country."[24]

Franklin was torn as to what course he should take. That his country would benefit from the emigration of skilled mechanics was obvious, but, as he advised his petitioners, neither Congress nor the states would consent to paying the cost of immigration, no matter how worthy the cause. Franklin knew nothing about the men other than that they were textile workers down on their luck. As one of them informed him, he had worked in Manchester for over twenty years, much of that time as a factory manager, but "this Wicked American War" had left him "scarcely able to get a Living."[25] Franklin advised the men to be prudent and wait until the war ended before making their move. Congress had cancelled his authority to issue the kind of pass they required.[26] But the Englishmen, mistaking Franklin's general words of encouragement for an ironclad vow of support, impetuously went ahead with their scheme, sold their assets, and tried to sneak out of the country. The luckless group ran afoul of revenue officers in Derry, Ireland, was thrown into jail, and after being released fortunately secured employment in the textile mills there.[27]

Sometime after this episode, and perhaps in part because of it, Franklin wrote a pamphlet for Europeans intending to sail for America. In it he warned his readers, as he had warned Wyld, that America, though a prosperous land, was not a place where fortunes waited to be seized. Even so, America could be a land of promise for the hardworking, for those industrious enough to make their own way. If Franklin hoped to discourage those with delusions of New World grandeur, he nonetheless realized that the new nation needed talented farmers and inventive mechanics.[28] His admonitions, then, did little to deter those like Wyld who looked to the new nation as a land of opportunity.

Other more fortunate English craftsmen slipped away under assumed names and from obscure ports. American manufacturers knew that those workers faced stiff fines and imprisonment if caught, but they recruited them without compunction. Some illegally emigrating British artisans had in fact introduced localized textile manufactures to the inland areas of South Carolina shortly after the end of the war. Emigrating Britons continued to evade officials well into the next century, bringing with them invaluable talents and a much-needed familiarity with new machines.[29]

Of course the British government could do nothing to restrict the emigration of other Europeans to the United States. One resident on the continent proposed a mass transatlantic transit of technology, a plan even more ambitious than that of Penet and Coulaux during the war. The Comte de Beaufort wrote the Continental Congress from Liège in October 1785, pointing out that since Europe was technologically rich and America technologically poor, the latter ought to make up the difference by learning from the former. He therefore proposed to cross over to the United States and "establish Arts & Manufactures . . . after the English and French modes." In return for a huge grant of land from the public domain somewhere along the Atlantic Coast, with rivers passing through it for water power, he would erect an entire manufacturing and farming community, with textile mills, iron forges, and furnaces for firearms and hardware, and various subsidiary operations. He would recruit the finest mechanics in Europe to work in the factories and the most talented farmers to introduce new techniques used in France and the low countries for hemp, flax, tobacco, wine grapes, and vegetables. He guaranteed that his miniature province would be more than self-sustaining, and he promised that it would support ten thousand families and boast a fleet of one hundred merchantmen by the end of twenty years.[30]

Congress, more than just a little interested in Beaufort's proposition, regretfully declined his offer. It had no land available to grant satisfying the Count's requirements for a salt-water port, fertile land, and running water for power. Another complication arose from Beaufort's unfamiliarity with American politics and apparent ignorance of Franklin's warning to prospective immigrants. He wanted to rule his domain as a semi-feudal duchy, an arrangement at odds with American republicanism.[31] Besides that, American shipping and manufacturing interests would have looked at the whole project with a jaundiced, jealous eye. Yet Congress's willingness to consider Beaufort's proposal showed how conscious politicians were of American technological immaturity and the need for radical steps to bring about change.

Many of the politicians in Congress at the time of Beaufort's petition were themselves actively seeking to uplift American manufactures and agriculture. As merchants and landholders they joined hands with the Coxes and Bordleys. Indeed, more and more Americans took an active interest in promoting large-scale technological growth.

George Washington became one of those promoters, a prime example of the devotee who advanced science and technology through manufactures and agriculture. After the War of Independence, Washington spoke out as an avowed proponent of home manufactures, both for export and for internal consumption. Early in 1789 he wrote Thomas Jefferson that "the introduction of the late-improved Machines to abridge labour, must be of almost infinite consequence to America."[32] Washington, then, was one of those who had become aware of the technological revolution underlying the Industrial Revolution in Great Britain. Like Tench Coxe, he felt that Americans "have already been too long subject to British prejudices"; in other words, Americans needed to achieve technological independence to guard their political independence.[33] Washington kept track of manufacturing experiments begun after the war, particularly a wool finishing operation at Hartford, Connecticut. He visited the Hartford factory while president and left disappointed because he had hoped the operation would be more impressive than it was. He was as impatient as Coxe. His vision of future American technological greatness was every bit as grand.

Washington's interest in manufactures was only part of a larger interest encompassing invention, internal improvements, animal husbandry, and, more than anything else, agriculture. Washington was still perhaps the most progressive farmer along the Virginia tidewater. Prominent British agronomist Arthur Young ranked him among the most scientific farmers in America.[34] After resigning as commander-in-chief in 1783, Washington retired to Mount Vernon and resumed the life of a gentleman farmer. He experimented with new crops, planting techniques, fertilizers, and machinery, as he had in the pre-war days. He complained that most farmers pursued "an unprofitable course of Crops, to the utter destruction of their lands." Lulled into thinking that ample western lands eliminated the need for judicious practices, they "pertinaciously adhered to" their ruinous habits.[35] Washington consequently believed that it was up to large landholders like himself to adopt new methods and set an example—just as Tench Coxe believed wealthy manufacturers should break the ground in their field. Both Washington and Coxe extended their patronage to inventors at a time when inventors badly needed encouragement.[36] They likewise shared a belief that government should do more than levy tariffs to promote American manufactures and technology in general. Washington's desire for a national experimental farm—like Coxe's bid for a national manufactory—though thwarted, was indicative of his concern for American technological progress.

By the early 1790s Washington had cultivated a wide-ranging circle of correspondents in Britain and the United States. His correspondence with Arthur Young was the most productive, with Young a willing participant in a by and large one-way, Britain to America, transit of technological information. Acting on his own, Washington, like Coxe and Bordley, helped build a community of Americans interested in promoting material progress through technological change. Equally as important, Washington lent his moral support to manufacturing and agricultural societies organized after the war.

Coxe and Bordley went even further and became actively involved in a number of these collective efforts. Some of the resulting "societies" were little better than clubs assembling once or twice a month to voice their backing of home manufactures, such as the numerous mechanics associations founded from Boston to Baltimore in the 1780s. Others, however, were more tightly organized and came up with practical manufacturing proposals.

The first of the post-war manufacturing societies was the Pennsylvania Society for the Encouragement of Manufactures and the Useful Arts, an organization built to some extent on the rubble of the United Company of Philadelphia. In 1787 the newly formed society cleaned out and restocked the old factory in Philadelphia and started over. It offered premiums (e.g., $20 for any laborsaving textile machinery), hunted for skilled artisans, and launched a publicity campaign to stir up interest. The society had trouble locating water-driven machinery—a commentary in itself on the state of American manufacturing technology, since Philadelphia was the center of post-war manufacturing activity through the 1780s. The difficulty of procuring suitable machines got the society off to an inauspicious start.[37] Apparently the only three available spinning jennies and one carding machine in Philadelphia had been whisked away by British agents, to the society's utter dismay. Local carpenter Enoch Richardson constructed new jennies, but the society had to go as far as Alexandria, Virginia, to find a mechanic skilled enough to make a carding machine. The need to build new machines in fact delayed the opening of the society's factory until spring 1788.[38]

Upon opening to the whir of newly made machines, the factory enjoyed moderate success for the first year of operation. Looms were added a bit later, giving it machines to weave as well as spin and card. The factory produced a few thousand yards of rough cloth, and members of the society manufacturing committee, one of whom was Tench Coxe, issued optimistic reports.[39] Their contagious enthusiasm spread to the newspapers and the public was so curious to see what was going on that the factory managers finally limited the number of visitors because of "Obstructions which the Persons" working at the jennies and looms met from "People crowding upon them."[40]

Despite an impressive output of coarse cloth during that first year, the factory began to lose money shortly after. The society itself slipped into a decline from which it did not recover. There was simply an insufficient demand for the society's coarse cloth, since cloth imported from Britain was cheaper, and the factory did not have the workers or the machines to produce fine quality cotton fabrics. What machines the factory did have were not as sophisticated as the Arkwright machinery used in British textile mills. In 1789 the society hunted fruitlessly for qualified mechanics to upgrade its operation and was soon forced to pay factory employees with cloth instead of money. The Pennsylvania General Assembly stepped in and bought one hundred shares of stock, bringing a temporary respite. But a fire destroyed the factory early the next year, and the society, unable to afford the cost of rebuilding, withered away.[41]

The Pennsylvania manufacturing society may have come too soon. Americans in the 1780s did not have the technological expertise to erect factories that could compete with British imports of coarse cloth, much less finer quality fabrics. True, there were more mechanics around the country familiar with textile machinery—dozens more than in pre-war years. After 1788 the Pennsylvania manufacturing society received more petitions than it could handle from mechanics offering to build jennies and looms.[42] Unfortunately, none of those mechanics was familiar with the most advanced Arkwright machinery. No models or detailed plans had as yet been smuggled into the country.

The experience of the Pennsylvania manufacturing society was representative of what happened to most other manufacturing societies through the end of the 1780s. There were perhaps twice as many manufacturing societies formed in the post-war as the pre-war years. They were not able, however, to bring sudden industrialization. Herbert Heaton observed that the post-war home manufactures "movement—one could almost say the crusade—was more remarkable for its enthusiasm than for its achievements."[43] Societies founded during the post-war "crusade" had a difficult time bringing in craftsmen conversant with the latest techniques used in Great Britain.[44] This, added to entrepreneurial inexperience, a shortage of capital, and inability to compete with British goods, caused a rise and fall among post-war societies like that of the pre-war societies. Americans may have been technologically inclined to the extent that they wanted the newest machines and could now identify them by name, but they were as yet unable to convert their desires into practical results. Some manufacturers had Hargreaves-type jennies and shuttle looms, but none had been able to combine these with Arkwright-type spinning machinery, much less Crompton mules, to escape bottlenecks, breakdowns, and capital loss. American manufacturing enthusiasts left the 1780s without having realized their ambitions.

They closed the decade in a frantic burst of activity. The end of the Revolutionary Era, marked by the inauguration of a new federal government under the Constitution, saw a virtual eruption of manufacturing companies around the country. The Constitution bolstered the manufacturing enthusiasts' sagging spirits. Artisans and mechanics marched proudly in processions staged in Boston, Philadelphia, and elsewhere around the country in 1788 that celebrated the ratification of the Constitution and the launching of a new federal ship of state. They hoped that the new government would breathe new life into the manufactures movement.

No sooner had ther Pennsylvania manufacturing society fallen on hard times than a club dedicated to "Mechanical Improvements" and a society for domestic manufactures organized in Germantown made their appearance.[45] At the same time, affluent New Yorkers joined in a cooperative effort to build a factory housing a carding machine, 2 jennies, 18 looms, and 140 spinning wheels.[46] An association of manufacturers and mechanics who pledged to further American industrial growth was formed in Providence, while Baltimore and Delaware residents established manufacturing societies of their own.[47] All believed that manufactures could be a social panacea. None weathered more than a few stormy years and for the same reasons as the Pennsylvania manufacturing society.

The Beverly cotton manufactory in Massachusetts, though holding on longer, also eventually succumbed. It had been the first of its type in New England after the war and second in the nation only to the Pennsylvania manufacturing society factory. Members of the Cabot family led the investors who combined to launch the Beverly experiment. Their biggest problem lay in obtaining suitable machinery. They erected a three-story factory in 1788 and managed to collar Thomas Somers, an immigrant English mechanic, to build the Arkwright-type machinery they needed. Somers, in fact, had been ready to leave the United States because of his failure to introduce textile manufactures in Maryland and Connecticut.[48] George Cabot enticed him to Bevery before he could leave, but Cabot and his fellow investors soon found that building machinery was both incredibly expensive and seemingly and unending, capital-intensive process. And Somers, disappointingly, was not able to duplicate Arkwright's designs. The Beverly manufactures exhausted their funds and appealed to the Massachusetts Assembly for aid, stating as their reason "the extraordinary price of machines unknown to our mechanics." The assembly agreed to loan them $1,000, and the factory eventually made some cloth; but it limped along, and the investors felt fortunate to break even. The Beverly group did not devise a fully integrated system and used machines that were largely handpowered.[49]

Of all the manufacturing projects begun within ten years of the end of the war, the cotton operation masterminded by Tench Coxe and Alexander Hamilton in Paterson, New Jersey, was the most ambitious. It too eventually failed. The ubiquitous Coxe, undeterred by the disappointments of the Pennsylvania manufacturing society, had originally hoped to involve the federal government directly in this new enterprise. Failing that, he, Hamilton, and a select group of financiers set out on their own and incorporated a Society for the Establishment of Useful Manufactures in November 1791. They were careful to recruit expert British mechanics, so the "SUM" escaped many of the machinery problems faced by earlier textile manufacturing experiments. By the time the SUM cotton factory at Paterson was ready for business in 1792, Arkwright machinery was being faithfully reproduced by local craftsmen such as William Pollard, a merchant-mechanic of Philadelphia. William Pearce,

Thomas Marshall, and George Parkinson, three talented British mechanics recruited at considerable cost, held down responsible positions in the SUM factory. Parkinson in particular was invaluable because he could modify Arkwright machinery to a variety of specifications.[50]

The SUM seemed destined for success. Money was no problem in the early months, talented mechanics had been set to work, and the appropriate machinery was installed. The directors of the society, however, soon proved to be too speculative for their own good. They invested in diverse and unprofitable ventures. They dreamed of erecting an economically self-contained community, a manufacturing town that would act as a model for the nation. Instead of sticking to cotton cloth, they dipped their hands into myriad projects, attempting too soon to make Paterson a factory town for paper, shoes, brass, iron, linen, and sail cloth. The directors diversified too quickly, before they had changed the buying habits of consumers who still preferred British over American textiles. And since the SUM was inseparably connected in the popular mind with Alexander Hamilton, it suffered a staggering blow when Hamilton emerged as leader of one of the two political parties forming by the mid-1790s. George Logan, a Pennsylvanian who had championed both manufactures and farming, lashed out at the SUM as an elitist enterprise designed to repress rather than uplift the masses. He and others who might have supported the SUM condemned it instead.

Succumbing to outside pressures and unwise investment policies in 1796, the SUM, having been "floated amid enthusiasm," finally closed its factory doors. Wealthy British clothier Henry Wansey noted on his American tour of 1794 that numerous other textile operations, despite their recruiting of skilled workers and new machines, became the victims of overambitious management like that of the SUM. This, added to the inferiority and high price of their cloth compared with that imported from Britain, was usually enough to spell their doom.[51]

The only textile manufacturing experiment to get its start in the closing days of the Revolutionary Era and to survive and flourish in the next decades was the famous Almy, Brown, and Slater undertaking in Pawtucket, Rhode Island. The SUM may have had an array of talented craftsmen and managers, but none could compare with Samuel Slater, the young Englishman hired by Moses Brown near the end of 1789 to help retool his Pawtucket factory, which he had already turned over to his son-in-law William Almy and nephew Smith Brown. Moses Brown was an imaginative entrepreneur, an experienced investor in candle and iron production as well as trade. Slater, for his part, had familiarized himself with Arkwright machine techniques and the details of factory management and work before leaving his native England. Under his expert guidance, local foundry owner Oziel Wilkinson (soon to become Slater's father-in-law), Wilkinson's son David, millwright Sylvanus Brown, and several immigrant English mechanics were able to make functional, water-driven Arkwright machinery. By the end of 1790 Slater had seen to the installation of drawing, carding, roving and spinning machines. Moses Brown, who continued to take an interest in the enterprise, had had jennies one or two years before Slater's arrival, but it took Slater, a capable manager (and soon after a partner with Almy and Brown), to devise an integrated system. Under the wise, realistic direction of Slater and his partners, the new factory did not take on too much too soon. During the first half dozen years it produced yarn only, which was sold through a central weave shop in Providence. Almy, Brown, and Slater kept their operations small, particularly when compared with the SUM. And while they initially used a converted fulling mill and later built a new factory house, they would keep one foot in the cottage industry. The combination of their modest ambitions and Slater's managerial skills kept the Pawtucket factory solvent while others came and went.[52]

Those struggling to hang on did not have an easy time of it. Connecticut entrepreneurs, for example, reported that by 1791 they had craftsmen turning out carding and spinning machines—so many, in fact, that they predicted they would soon close the technological gap with Britain. Yet mixed in with this boastful enthusiasm were the complaints that Americans did not raise enough sheep for wool, that they did not import enough cotton from the West Indies, and that British imports continued to dominate the market.[53] If their operations seemed to limp along, so did those of textile entrepreneurs nationwide.

Other industries also went through hard times in the 1780s, with the iron industry especially suffering a minor depression. Some bar iron was exported, but most was consumed at home, primarily because the British iron industry had expanded and improved, thereby lessening the need for imports. When collecting data for his *Report on Manufactures,* Alexander Hamilton had received optimistic reports on iron production in Connecticut and Virginia, but iron production in both states was small-scale and competitive only in local markets.[54] Most American ironmasters, unlike the rising class of textile manufacturers, "resisted innovation" and held on to "old practices," using charcoal instead of coal for smelting and water-driven machinery, which limited the size of their shops.[55] Imported ironware therefore competed on a par with American-made goods, and most steel came from Great Britain. Iron manufacturers only slowly changed their operations and converted to coking, puddling, and steam power, well behind the British. Britain held onto its technological secrets in the iron industry more successfully than it had in textiles, but then the American ironmasters were not all that enthusiastic about making expensive changes anyway.

Americans during the 1780s pushed for improved agriculture with one hand while pushing for improved manufactures with the other. Unlike

14. Samuel Slater, the immigrant adopted as native son. Courtesy of the Library of Congress.

the pre-war years, some post-war associations showed increasing specialization, dividing into agricultural and manufacturing societies. Agriculturalists and industrialists were not consciously moving apart, however. On the contrary, both still worked toward what they saw as the same end. Division, at this stage at least, cannot be attributed to divergent interests. Tench Coxe had not been disingenuous in arguing that agriculture was as important as manufactures. The farmer helped the industrialist by producing the cotton, wool, help, and flax for his machines and vegetables, fruit, and livestock to feed his workers. The industrialist helped the farmer by expanding the market for agricultural produce and supplying credit. More often than not, magazine and newspaper articles written to promote manufactures included lines on agriculture. In many cases the industrialist and landholder were one and the same, an extension of the merchant-farmer connection so central to colonial economic life.[56]

The Philadelphia Society for Promoting Agriculture, organized in February 1785, numbered among its twenty-three charter members Benjamin Rush, Richard Wells, Samuel Powel, George Logan, and John Beale Bordley. Rush's concern for manufactures dated back to the 1760s; Richard Wells, inventor and merchant, had similar interests; Samuel Powel served as vice-president of the Pennsylvania manufacturing society concurrently with his term as president of the Philadelphia Agricultural Society; George Logan, later to become a staunch Jeffersonian and critic of Hamilton and the SUM, crusaded for home manufactures and served as president of the Germantown manufacturing society in 1790; John Beale Bordley later became a devoted Hamiltonian, but he recognized no distinction between agriculture and industry when it came to promoting the national welfare.[57] As further proof of the widespread linking of agriculture and industry, a New Jersey society devoted to progressive farming founded in 1790 amended its constitution to include the promotion of domestic manufactures. Tench Coxe, a member of the Philadelphia Agricultural Society, may have had a hand in this conversion.[58]

Agricultural societies aimed to import and improve on the latest technology from Europe, just like manufacturing societies. They had a slight advantage in not having to resort to subterfuge and espionage to obtain the data they wanted. Between 1785 and 1795 agricultural societies were founded in most states from Georgia to Massachusetts. Like the manufacturing societies, they displayed a sense of inferiority to Britain and a burning desire to turn the tables. One magazine, commenting on the founding of the Philadelphia society, expressed its wish that that society, patterned after successful organizations in Europe, would enable the United States to seize the lead.[59] Indeed, Philadelphia Agricultural Society members admitted, "It was a conviction of our great inferiority in this respect, which gave rise to the present Society."[60]

Despite the high hopes attending the founding of these societies, they, like their manufacturing counterparts, did not bring the quick results that their members expected. Their advantage over the industrialists turned out to be more apparent than real. Ten years after the formation of the first societies in 1785, not that much had changed, regardless of the premiums awarded and the papers published. The Massachusetts Society for Promoting Agriculture, founded in 1793, had the same general goal as the Pennsylvania society eight years before: to correct "the imperfect state of husbandry" by securing models of machines invented by Europeans and copies of their pamphlets and articles.[61] Chancellor Robert Livingston, speaking as president of a New York society for agriculture and manufactures established in 1791, pleaded with his listeners to close the technological gap between Europe and America. Once Americans learned the secrets of progressive agriculture, he contended, nothing could stop them from jumping to the world lead because of their endless reserve of fertile land.[62]

But the agriculturalists, like the industrialists, discovered that mastering those "secrets" was not easy, and getting the average farmer to change his ways was even more difficult. As in the 1760s and 1770s, progressive farming in the 1780s and 1790s was not a nationwide phenomenon. Only the elites, the Washingtons and Livingstons, had the time, resources, and commitment to bettering American agricultural life. Reform drives and plans for state-sponsored farms along the lines proposed by the Philadelphia agricultural society in 1794 fizzled out, defeated by an unresponsive public. As Washington complained to Arthur Young, where land is cheap and labor dear, farmers are more likely to cultivate much, rather than cultivate well.[63]

Successful technological change in agriculture and manufactures was not easy to bring about, and investment in either one based solely on enthusiasm or a sense of national duty was never long-lived. Most farmers innovated only when it was profitable to them. With good reason had Franklin noted that "it is a Folly to think of forcing Nature."[64] They may have wanted the new nation to become technologically advanced, but not at their expense. Most manufacturers were equally circumspect. Grueling commercial competition eliminated the faint-hearted and the overspeculative. Only the careful entrepreneur teaming machines, skilled labor, sound management, and select production survived and maintained his incentive to be technologically innovative.

In the 1780s American manufacturers, except those attached directly to commerce and shipbuilding, were still not competitve enough to pose an immediate threat to Great Britain. The British feared the manufacturing potential of the United States and continued to restrict the flow of technological information, but that potential was not immediately realized. After having viewed the SUM plant in Paterson, British minister

George Hammond cautioned that "no small degree of vigilance will be requisite in Great Britain to prevent the emigration of artists and the export of models of machines."[65] Hammond's fear notwithstanding, the American economy managed as well as it did only because of the reinvigoration of the carrying trade and export of agricultural surpluses, not because of manufacturing.[66]

Even the most cautious American manufacturers enjoyed no more than middling success through the end of the eighteenth century. This should not be too surprising, considering the relative youth of the country and its distance, politically and geographically, from Britain, the center of industrial progress. American manufacturers had to compete against long-established and more adept British entrepreneurs, a risky business at best. Almy, Brown, and Slater were doing moderately well in Pawtucket, yet alone they did not threaten the British lead in textiles. They had not triggered a sudden conversion to factory work, nor had they brought incipient mass production. Indeed, the "U.S. cotton spinning industry lay far behind that of Britain in 1790"; the British were not suddenly eclipsed.[67]

Despite the lack of visible returns for the present, the stage had been set for greater success in the future. The post-war surge for domestic manufactures did not bring immediate dividends, yet it did bring a taste of what could be expected. The new nation had committed itself to developing an indigenous technological expertise. It was that commitment that George Hammond had detected and feared most.

The success of Almy, Brown, and Slater in Pawtucket gave credence to Hammond's fear. They had done something that had not been done before. They had proved that, given time, the new nation could industrialize. Americans had begun to pull together the elements of a new technological order that would satisfy their national ambitions, and the firm of Almy and Brown, backed by Moses Brown and later led by Samuel Slater, showed the way. They had the raw materials—cotton for cloth, wood for machines and fuel, water for power; they had the inventive craftsmen; they had the machines; they had the capital; they had the entrepreneurial skills; and they knew what the market would bear. The political revolution of the 1780s brought the political climate they wanted, both in the achievement of independence and the later adoption of the Constitution. Their Pawtucket operation is most significant as a symbolic gesture of technological independence, a harbinger of the industrial transformation to come.

Phineas Bond, British consul for the Middle states during the post-war years, had noticed that the Americans were moving ahead technologically. He took pains to comment on the status of American manufactures in a report to the home office in 1788. Referring to the numerous textile experiments of the decade, he reported that "they are essentially deficient in those main sinews of advancement, money, artificers, and fit utensils; still their exertions are made with great zeal and the improvements tho' small are progressive."[68] In more prosaic terms, Bond meant that if the Americans trailed behind the mother country, it was not for lack of trying to overtake her lead. They were determined to catch up.

They did not do so immediately, though they would eventually. American society was changing. In the colonial period provincial assemblies had had to act extralegally to support most manufacturing experiments. Banking and the granting of corporate charters were discouraged where they were not outlawed altogether. In short, the politics of empire worked against large-scale technological change. Manufacturers after the War of Independence did not encounter the same problems. State governments encouraged incorporation and granted bank charters. Legislatures issuing charters to banks, canal companies, and manufacturing societies tried to promote public and private interests simultaneously. They often awarded charters to private investors if they thought those investors would perform a public service. The national government organized under the Constitution gave further encouragement. For longtime advocates of technological change like Dr. Benjamin Rush, the Constitution was essential because it could be used to advance manufacturing, agriculture, science, and the "arts." Where government had often been an obstacle to technological change in the colonial years, in the new nation it became an ally.

The federal government provided under the Constitution assisted in the race to catch up with the British by levying tariffs and fixing monetary policy. It nearly outstripped Parliament in its support for manufactures, as did state governments through tax exemptions and bounties. The growth of private banking, the founding of the Bank of the United States in 1791, and the establishment of a national mint in 1794 (providing for a uniform currency) helped build a sounder credit structure and a more fluid economy. Potential investors in manufactures became more confident as a result.[69] They eventually took some of the capital that they had accumulated in the carrying trade and transferred it to manufactures.

In a precedent-setting series of steps, the first Congress that was formed under the Constitution in 1789 defined the limits to its promotion of science and technology. Naturally some manufacturers had hoped that the new federal government would use its broad powers to go beyond the economic interventionism implicit in taxes and import and tonnage duties to foster directly the growth of individual industries. John Amelung of New Bremen, Maryland, became the first to test the technological limits of Congressional support. Amelung owned a glassmaking operation, which he had opened in 1785 and which he had filled with skilled craftsmen recruited in Europe. His friend and fellow Marylander Daniel Carroll recommended that since Amelung had done well until a fire swept through his works, Congress ought to loan him the money to rebuild. A number of congressmen concurred with Carroll, taking the position that Congress is "vested with a general power to encourage the arts and manufactures," while others objected that the government could not legally make loans to individuals. Legal considerations aside, opponents contended that such "piecemeal" aid would not help the economy all that much and would serve to fritter away monies that Congress could not afford to lose.

In the end the opponents to Amelung's petition won the debate, and the loan was refused.[70] Congress, by so doing, drew a line it would not cross at that point. During the first two sessions of Congress, a week rarely passed without a petition being read into the record from merchants or mechanics in Boston, Philadelphia, or some other town requesting economic intervention. Some advocated higher duties on imports, some wanted lower, some wanted more duties, some clamored for fewer. Still others pled for the type of assistance denied Amelung. All were turned away.

Congress chose to go the middle route, adopting a paternalistic form of laissez-faire. Though Amelung did not get his loan, Congress did pass an import duty on glass to protect him and other glassmakers. Likewise, most congressmen saw to it that basic industries in their districts, from wool card making to shipbuilding, enjoyed some sort of protection. In theory this protectionism, by saving Americans from voracious European manufacturers, by allowing them to compete as equals at home while being shielded from the "wolves" from without, defended the principles of free trade. In practice, the result was to institutionalize government support of American technological growth. Government economic interventionism was an extension of that support. Hamilton's *Report on the Subject of Manufactures* simply called for a larger dose of a medicine already in use, a continuation of a treatment long dedicated to promoting national technological health.

Though the national government avoided handing out subsidies to Amelung in 1790, it came close to reversing itself less than a decade later with Eli Whitney. Whitney gained a sympathetic ear and a generous contract in 1798 because he promised to turn out muskets by the thousand, using more machine power and less human labor. The government backed Whitney in his venture and put up much of the capital he needed to build his Mill Rock plant. Not only did the government invest in Whitney's venture, it underwrote the machine tool experiments and inventions perfected at the Springfield and Harpers Ferry armories. As a result it became involved directly with the promotion of invention and technological innovation.[71]

Government involvement notwithstanding, it would be decades before the United States became more of a technological "initiator" than a technological "borrower." Yet the desire to bring on that change, and the nascent vision of how it could be accomplished, dated back to the 1790s.[72] In a series of essays published in the *American Museum* in 1792, Tench Coxe slipped on the robe of prophet to tell the story of future American technological greatness. He launched into a lengthy point-by-point rebuttal to Lord Sheffield, who nearly ten years earlier had jeered that political independence condemned Americans to a sedentary life, a perpetual subservience to British manufacturing and commerce. Ticking off one advance after the next in agriculture and industry, Coxe countered that the United States had progressed, not regressed, and each year since the war had brought, and would continue to bring, more signs of success. Perhaps recalling the failures of wartime munitions production, Coxe noted that twenty-one gunpowder mills had been built since 1775, and more were under construction—a thinly veiled threat that the next war would find Americans better prepared and capable of taking care of themselves. Admitting that his countrymen still imported over one third of their iron products, he predicted that the United States would eventually take away that third and cut into other markets long monopolized by the British. Coxe, as developments over the next half century would show, was the better prophet. Independence was a technological beginning, not an end.[73]

The post-war years can be viewed as a gestation period preceding the birth of a technological America foretold by Tench Coxe. In effect, those years became the intellectual equivalent to Walt Rostow's economic precondition to takeoff. During the 1780s American appreciation for and identification with technology took shape, setting the stage for changes in the next century that propelled the nation into the industrial age.

Naturally all Americans did not suddenly and uniformly become scientific farmers or industrial innovators. The average farmer in Revolutionary America depleted more soil than he replenished and made use of only the most basic tools. From his perspective, he was making effective use of limited capital. Similarly, most iron mill operators resisted changes that increased their costs. Adventurous entrepreneurs like Andrew Ellicott invested in the steamboat and automated milling experiments of the 1780s, yet many others remained aloof. While some merchant-manufacturers like Moses Brown risked their capital in new, more ambitious ventures, others played it safe and kept their money in the carrying trade or smaller manufacturing operations. There would always be some distance between belief in technology as a source of progress and the translation of that belief into practice.[74] As one would expect, market pressures often dictated the rate of technological change. But of utmost importance, technological innovation by 1790 had become tangible enough for Americans to see how the translation of belief into practice could be accomplished. As far as many Americans were concerned, there was a direct connection between national prosperity and the rate of technological change. They treated the political republic and the technological republic as obverse sides of the same coin.

NOTES

1. *Freeman's Journal*, 14 November 1781; supposedly a letter from Silas Deane to Robert Morris. Even if the authorship is spurious, the contents are on target.

2. See for example the *Pa. Gazette*, 14 September 1785 and 25 June 1788; the *American Magazine* 1 (June 1788): 439-440; and the *Gentleman and Lady's Town and Country Magazine* (May 1789):202 and (August 1789):280. Also see Hugo A. Meier, "American Technology in the Nineteenth Century," *American Quarterly* 10 (1958):116-130. Thomas C. Cochran has noted that "After the Revolution, American enthusiasm for mechanical innovation that would increase domestic production appears to have been more widespread than historians have generally understood." See his *Frontiers of Change* (New York: Oxford University Press, 1981), p. 53. Cochran emphasizes the cultural (more than the purely geographical) sources of technological change.

3. From William Barton's "Essay on the promotion of American manufactures," *American Museum* 2 (September 1787):257-260.

4. *Pa . Gazette*, 5 October 1785.

5. Jay in the PCC, Item 71, II, 387-395 (reel 85); and Barton, *The True Interest of the United States and Particularly of Pennsylvania Considered* (Philadelphia: Matthew Carey, 1786), pp. 27-28.

6. Homespun, "Plain Thoughts on Home Manufactures," *Columbian Magazine* 1 (February 1787):281-284; also Rezneck, "The Rise of Industrial Consciousness," pp. 788-789; Bell, "Science and Humanity in Philadelphia," pp. 35-36, 51-63; Kasson, *Civilizing the Machine*, pp. 14-36, for the linking of technology with a republican creed; and Charles L. Sanford, *The Quest for Paradise* (Urbana: University of Illinois Press, 1961), pp. 155-175, which makes a burst of insightful observations about American republicanism and industrialization, all within the context of Sanford's larger argument about the pursuit of a New World moral order. For an excellent discussion of post-war technological changes, see Cochran, *Frontiers of Change*, pp. 50-77. McCoy, *The Elusive Republic*, pp. 105-119, contends that Americans had to make an ideological shift in the 1780s to embrace large-scale manufactures. I can agree with McCoy only in part. A cautious few did advise against large-scale manufactures (e.g., Franklin or admonitions like those in *Pa. Journal*, 7 February 1784). Yet those objections were not so much to bigness as they were to what the writers thought bigness represented in Britain: exploitation and aggrandizement. The combination of their inventive curiosity with their desire to become more competitive and their confidence in their own moral superiority helped to counter their fears. Besides, Americans had taken to condemning the British for their supposed moral decline long before the beginnings of industrialization. Thus those like George Logan who objected to Hamilton's program were opposed to the context that Hamilton put manufactures in, and perhaps their scale, but not to their existence. Those who accepted small-scale manufacturing opened the door for still larger enterprises. For others

of the Revolutionary generation, factories were not tainted in and of themselves; rather, their social benefit or danger depended on how they were used, and in America, many felt, they would serve a higher purpose.

7. Jeremy, "British Textile Transmission to the U.S.," pp. 24-25; and Nathan Rosenberg, *Technology and American Economic Growth* (New York: Harper and Row, 1972), p. 25.

8. Butterfield, ed., *Letters of Benjamin Rush*, I, 553.

9. *Pa. Gazette*, 30 April 1788. Americans did not, however, simply imitate the British. For example, as Louis C. Hunter noted in "Waterpower in the Century of the Steam Engine," in Hindle, ed., *America's Wooden Age*, pp. 160-192, Americans used water power to drive their mills more often than they used steam at least through 1860. The primary importance of Watt to the American obsession with mechanization was his symbolic role in having fostered industrialization in Britain. Americans looked to Boulton and Watt for leadership in only the most general sense. When it came to building their own factories, they turned to water more often than steam power because it was a more economical source—just as British industrialists, lacking that cheap supply, turned *somewhat* earlier to steam. As Dolores Greenberg has pointed out in her "Reassessing the Power Pattern of the Industrial Revolution: An Anglo-American Comparison," *American Historical Review* 87 (1982):1237-1261, the British too used water power on a significant scale even after 1860. Therefore generalizations about the rate of conversion of power sources—like generalizations about the Industrial Revolution itself—have to be made very carefully.

10. Brooke Hindle, *Emulation and Invention* (New York: New York University Press, 1981), p. 1.

11. Tench Coxe, *An Address to the Assembly of the Friends of American Manufactures* (Philadelphia: Robert Aitken and Sons, 1787), p. 11.

12. Idem, *Observations on the Agriculture, Manufactures and Commerce of the United States* (New York: Francis Childs and John Swaine, 1789), pp. 33-34. Also see Jacob E. Cooke, *Tench Coxe and the Early Republic* (Chapel Hill: University of North Carolina Press, 1978), pp. 98-108.

13. Bagnall, *Textile Industries*, I, 75; Bishop, *Amer. Manufactures*, I, 397; a copy of the contract between Coxe and his agent in the enterprise is in the Papers of Tench Coxe in the Coxe Family Papers in the Historical Society of Pennsylvania (microfilm publication), Philadelphia, 1977, series II, Correspondence and General Papers (reel 50), dated 9 August 1787.

14. *Conn. Hist. Soc. Coll.*, XXIII, 173, 197.

15. Letter from Mitchell to Coxe of 4 June 1788, in the Coxe Papers, Series II (reel 52), HSP, film copy.

16. Ashton, *Industrial Revolution*, p. 50.

17. Mitchell to Coxe, 4 June 1788, in the Coxe Papers, series II, (reel 52), HSP MSS.; also Richard Coxe to Tench Coxe, 9 March 1790 in ibid., (reel 55). For reflections on the competitiveness of machinery importation and a useful overview of the post-war quest for mechanical devices, see Anthony F. C. Wallace and David Jeremy, "William Pollard and the Arkwright Patents," *WMQ*, 3rd series, 34 (1977):404-425.

18. Metcalf Bowler, *A Treatise on Agriculture and Practical Husbandry* (Providence: Bennett Wheeler, 1786), pp. 9-10.

19. John Beale Bordley, *A Summary View of the Course of Crops in the husbandry*

of England & Maryland (Philadelphia: Styner & Cist, 1784), pp. 3-8; also see Olive M. Gambrill, "John Beale Bordley and the Early Years of the Philadelphia Agricultural Society," *PMHB* 66 (1942):410-439.

20. Charles Varlo, *A New System of Husbandry*, 2 vols. (Philadelphia: n.p., 1785).

21. Published in the *Gentleman and Lady's Town and Country Magazine* for July 1784, pp. 101-103, and successive issues through October 1784.

22. *American Magazine* 1 (October 1788):817.

23. See his *Essays on Agriculture* (New York: Francis Childs and John Swaine, 1790).

24. The letters sent by Henry Wyld and Edmund Clegg (another member of the group) are in the Franklin Papers, APS MSS. See in particular, Wyld to Franklin, 15 January and 12 February 1782; and Clegg to Franklin, 4 April 1782, in XXIV, fos. 141 and 76; and XXV, fo. 6, respectively.

25. Clegg to Franklin, 24 April 1782, XXV, fo. 39, APS MSS.

26. Franklin to Henry Royle et al., 4 January 1782, in Smyth, ed., *Writings of Franklin*, VIII, 355-356; and Franklin to Wyld, 31 March 1782, ibid., VIII, 410-411.

27. David Young to Sir Edward Newenham, 13 December 1782, in the Franklin Papers, XLVIII, fo. 5, APS MSS.

28. Benjamin Franklin, "Information to Those Who Would Remove to America," in Smyth, ed., *Writings of Franklin*, VIII, 603-614.

29. For South Carolina see the *American Museum* 8 (October 1790), appendix, p. 11. For emigration in general see Herbert Heaton, "The Industrial Immigrant in the United States, 1783-1812," *APS Proceedings* 95 (1951):519-527; Rowland Berthoff, *British Immigrants in Industrial America, 1790-1950* (Cambridge: Harvard University Press, 1953); and David J. Jeremy, *Transatlantic Industrial Revolution* (Cambridge: MIT Press, 1981), pp. 76-91, 252-257. Also see a proposal submitted to Alexander Hamilton in 1791, suggesting that the government actively recruit British artisans in Arthur H. Cole, ed., *The Industrial and Commercial Correspondence of Alexander Hamilton* (Chicago: A. W. Shaw Co., 1976), pp. 109-112.

30. Beaufort's proposal is in the PCC, Item 78, VIII, 151-174 (reel 94); original French and English translation.

31. Ibid., Item 80, 23 (reel 196); Ford, ed., *JCC*, XXX, 53. Congress also turned down a more modest proposal from Irishman William McCormick (PCC, Item 41, VI, 397, reel 50; Ford, ed., *JCC*, XXVIII, 221).

32. Letter of 13 February 1789 in Fitzpatrick, ed., *Writings of Washington*, XXX, 199.

33. Ibid., XXX, 186-187, letter to Lafayette of 29 January 1789.

34. Letter of 7 January 1786 in the GWP, series 4 (reel 95); also see the letter of 19 May 1789 in ibid., series 4 (reel 98).

35. Fitzpatrick, ed., *Writings of Washington*, XXVIII, 200-201, 511.

36. Washington's connection with James Rumsey and involvement in the Fitch-Rumsey steamboat controversy is a familiar story and is alluded to in the next chapter. In 1787 Coxe entered into a partnership with Robert Leslie, a Philadelphia clockmaker and inventor (and friend of Oliver Evans), whereby in return for a seven-year monopoly on the lead processing machine made by Leslie, Coxe in effect bankrolled the project. See the Tench Coxe Papers, series

II (reel 50), the agreement of 20 August 1787, HSP MSS. According to the *Pa. Gazette*, 27 May 1785, Leslie had earlier gathered "every model, drawing or description of any machine implement or tool" used in foreign countries for manufacturing. According to the *American Magazine* (March 1788), p. 271, Leslie also invented new types of clock pendulums.

37. See Jeremy, "British Textile Transmission to the U.S.," pp. 29-32; Leonard Bernstein, "The Working People of Philadelphia from Colonial Times to the General Strike of 1835," *PMHB* 74 (1950):322; Cooke, *Tench Coxe*, pp. 102-108; and a notice of the society's constitution is in the Tench Coxe Papers, series II, (reel 50), under August 1787, HSP MSS. Hunting for artisans is noted in a letter of Phineas Bond to Lord Carmarthen of 20 November 1787 in the *American Historical Association Annual Report* (1896):553.

38. Meeting of the Manufacturing Committee, minutes, 20 September 1787; 19 and 22 January 1788, Pennsylvania Society for the Encouragement of Manufactures and the Useful Arts, I, HSP MSS. Also see Heaton, "The Industrial Immigrant," pp. 525-526; and Joanne Neel, *Phineas Bond* (Philadelphia: University of Pennsylvania Press, 1968), pp. 55-60.

39. "The Board of Managers of the Penn. Society for promoting Manufactures and the Useful Arts, The Report of the Committee for Manufactures," *Columbian Magazine* 2 (December 1788):736-740.

40. *Pa. Gazette*, 19 March 1788; meeting of the Manufacturing Committee, minutes, 30 April 1788, Pa. Soc. for Manufactures and Useful Arts, I, HSP MSS.

41. Meetings of the Manufacturing Committee, minutes, 29 April and 21 May 1879, Pa. Soc. for Manufacturers and Useful Arts, II, HSP MSS. Also Hazard, ed., *Pa. Archives*, 1st series, XI, 569; and the general accounts in Bagnall, *Textile Industries*, I, 77-79; and Bishop, *Amer. Manufactures*, I, 405-410.

42. See the petitions of James Barbour and Robert Lemmon in meetings of the Manufacturing Committee, minutes, 2 and 20 February; 12 March 1788, Pa. Soc. for Manufacturers and Useful Arts, II, HSP MSS. Also see the letter from John Campbell of York, Pa., to Tench Coxe of 9 December 1788 in the Tench Coxe Papers, series II (reel 52), HSP MSS.

43. Heaton, "The Industrial Immigrant," p. 522.

44. Jeremy, "British Textile Transmission to the U.S.," p. 43, pointed out that this was the case in Philadelphia; there is no reason to suspect that other cities fared better.

45. See the 1790 broadsides of the *Constitution and Fundamental Rules of the Pennsylvania Society for Mechanical Improvements* (Philadelphia: n.p., 1790); and *The Constitution of the Germantown Society for Promoting Manufactures* (Philadelphia: Eleazer Oswald, 1790); for the bumper crop of post-war manufacturing societies, see Bagnall, *Textile Industries*, I, 89-134. Also see Whitfield Bell, "The Federal Processions of 1788," *New York Historical Society Quarterly* 46 (1962):5-39.

46. *American Museum* 7 (June 1790), appendix IV, p. 27.

47. *The Charter, Constitution and Bye-Laws of the Providence Association of Mechanics and Manufacturers* (Providence: Bennett Wheeler, 1789); Delaware and Baltimore in the *American Museum* 5 (February 1789):174-175 and (June 1789):59, respectively.

48. *Pa. Gazette*, 28 May 1788.

49. The quote is from George S. White, *Memoir of Samuel Slater* (Philadelphia:

n.p., 1836), pp. 54-55; also see Bishop, *Amer. Manufactures*, I, 398-399; Bagnall, *Textile Industries*, I, 89-98; and Robert Lovett, "The Beverly Cotton Manufactory or Some New Light on an Early Cotton Mill," *Business Historical Society Bulletin* 26 (1952):218-237. The Massachusetts government had earlier advanced £200 to Hugh Orr of Bridgewater to defray the costs of a jenny and carding machine being made by two Scottish brothers, recent immigrants to the state.

50. See Harold Syrett, ed., *The Papers of Alexander Hamilton*, 25 vols. (New York: Columbia University Press, 1961-1977), X, 345; for Pearce in particular, see Carroll W. Pursell, Jr., "Thomas Digges and William Pearce: An Example of the Transit of Technology," *WMQ*, 3rd series, 21 (1964):551-560. Parkinson, an expert on flax who eventually patented a flaxdressing machine in 1791, had contacted Benjamin Franklin back in 1789. See his letter to Franklin of 22 December 1789 in the Franklin Papers, XXXVI, fo. 191, APS MSS. Wallace and Jeremy, "William Pollard and the Arkwright Patent," p. 407, point out that Pollard had emigrated to America before he learned about textile machinery, but he employed in his shop English mechanics who had crossed over only recently.

51. Joseph S. Davis, *Essays in the Earlier History of American Corporations*, 2 vols. (Cambridge: Harvard University Press, 1917), I, 349-522; and the prospectus for the SUM in Syrett, ed., *Papers of Hamilton*, IX, 144-153. Also see the collection of documents in Cole, ed., *Industrial and Commercial Correspondence of Alexander Hamilton*, pp. 183-228; and Jacob E. Cooke, "Tench Coxe, Alexander Hamilton, and the Encouragement of American Manufactures," *WMQ*, 3rd series, 32 (1975):369-392. Also see David J. Jeremy, ed., *Henry Wansey and His American Journal* (Philadelphia: American Philosophical Society, 1970), pp. 40, 68, 82-83.

52. See David J. Jeremy, "Innovation in American Textile Technology During the 19th Century," *Technology and Culture* 14 (1973):40-76; Bagnall, *Textile Industries*, I, 84-165; Paul E. Rivard, "Textile Experiments in Rhode Island, 1788-1789," *Rhode Island History* 33 (1974):35-45; and Gary B. Kulik, "The Beginnings of the Industrial Revolution in America: Pawtucket, Rhode Island, 1672-1829" (Ph.D. dissertation, Brown University, 1980), pp. 116-162.

53. Cole, ed., *Industrial Correspondence of Hamilton*, pp. 10, 16-17, 36.

54. Ibid., pp. 101-102, 229-230.

55. Pearse, *History of Iron Manufacture*, pp. 129-130; Nettles, *Emergence of a National Economy*, pp. 263-291. The 14 December 1785 issue of the *Pa. Gazette* carried an advertisement by John Nancarrow for his "blistered steel," but the advertisement directly across from it was for imported British iron and steel tools.

56. Noted by Bell, "Science and Humanity in Philadelphia," p. 50, and evident in most essays written in the 1780s and 1790s. See for example Enoch Edwards's "Address to the Philadelphia county society for the promotion of agriculture and domestic manufactures of February 2, 1789," in the *American Museum* 5 (June 1789):554-559; or William Barton's article in ibid., 7 (May 1790):285-292.

57. See Frederick B. Tolles, "George Logan and the Agricultural Revolution," *APS Proceedings* 95 (1951):589-596; and idem, *George Logan of Philadelphia* (New York: Oxford University Press, 1953), p. 119. Also see the plea of AGRICOLA (probably Logan) in the *Pa. Gazette*, 23 April 1788; Gambrill, "John Beale Bordley," passim; and Lucius F. Ellsworth, "The Philadelphia Society for the Promotion

of Agriculture and Agricultural Reform, 1785-1793," *Agricultural History* 42 (1968):189-199.

58. The long arm of Tench Coxe can be seen here. See the letters from William Coxe to Tench Coxe of 10 and 30 January 1790, in the Tench Coxe Papers, series II (reel 55), HSP MSS.

59. *Boston Magazine* 2 (March 1785):113.

60. *An Address from the Philadelphia Society for Promoting Agriculture* (Philadelphia: n.p., 1785); also see *An Address to the Public from the South Carolina Society for Promoting and Improving Agriculture and Other Rural Concerns* (Charleston: A. Timothy, 1785).

61. See the Massachusetts society's *Laws and Regulations* (Boston: Isaiah Thomas and Ebenezer T. Andrews, 1793), pp. iii-iv. For the post-war agricultural movement in general, see Hindle, *Pursuit of Science*, pp. 355-367; and Nettles, *Emergence of a National Economy*, pp. 243-251.

62. Livingston's presidential address is in the *Transactions of the New York Society for the Promotion of Agriculture, Arts, and Manufactures*, part 2 (New York: Childs and Swaine, 1794); see also Donald Marti, "Early Agricultural Societies in New York: The Foundation of Improvement," *New York History* 48 (1967):313-331.

63. Washington's letters to Young of 5 December 1791 and 16 June 1792 are in Fitzpatrick, ed., *Writings of Washington*, XXXI, 437-440; and XXXII, 67-68, respectively. Also see Margaret Rossiter, "The Organization of Agricultural Improvement in the United States, 1785-1865," in Oleson and Brown, eds., *The Pursuit of Knowledge in the Early Republic*, pp. 279-298; and Rodney H. True, "The Early Development of Agricultural Societies in the United States," *Amer. Hist. Assn. Annual Report* (1920):295-306, for the continuing British influence. The Philadelphia agricultural society's unrealized proposal was published as *Outlines of a Plan for Establishing a State of Agriculture in Pennsylvania* (Philadelphia: Charles Cist, 1794).

64. Franklin, "Information to Those Who Would Remove," p. 611; also see North, *The Economic Growth of the United States, 1790-1860*, p. 8.

65. As cited in Heaton, "The Industrial Immigrant," p. 523.

66. See the gloomy post-war economic picture in Nettles, *Emergence of a National Economy*, pp. 45-64; the brighter picture in Merrill Jensen, *The New Nation* (New York: Alfred A. Knopf, 1950), pp. 177-257; and the more balanced assessment in Gordon Bjork, "The Weaning of the American Economy: Independence, Market Changes, and Economic Development," *Journal of Economic History* 24 (1964):541-560, which is not to say that the "spillover" profits from commerce did not indirectly promote manufactures; cf. Walton and Shepherd, *The Economic Rise of Early America*, pp. 1-4. Nevertheless, Neel, *Phineas Bond*, p. 78, cites a January 1791 report of the Board of Trade stating confidently that Americans were no threat because "it is astonishing how much they prefer agriculture to manufacturing."

67. Jeremy, *Transatlantic Industrial Revolution*, p. 19.

68. Letter of Bond to Lord Carmarthen of 16 November 1788 in the *Amer. Hist. Assn. Annual Reports* (1896):581.

69. Nettles, *Emergence of a National Economy*, pp. 112-121; Clark, *History of Manufactures*, I, 229-230.

70. *Annals*, 1st Congress, 2nd session, 1686-1688.

71. Harold C. Livesay, *American Made* (Boston: Little, Brown and Co., 1979), pp. 19-50.

72. See Rosenberg, *Technology and American Economic Growth*, pp. 59-86, for America as "Borrower," and pp. 87-116 as "Initiator." Also see idem, "American Technology: Imported or Indigenous?" *American Economic Review* 67 (1977):21-26. In a letter to the Royal Society of Arts of 2 April 1801, American James Meese noted that his country still had much to learn from Britain. "In agriculture, I may truly say, we are miserably deficient." His letter is in Great Britain, Series B, Royal Society of Arts, Loose Archives, A 16/6 (reel 2), LC microfilm. Also see A. E. Musson, "British Origins," in Otto Mayr and Robert C. Post, eds., *Yankee Enterprise* (Washington, D.C.: Smithsonian Institution Press, 1981), pp. 25-48, which argues that the American system of manufactures was in many ways derivative, not original, borrowing heavily from Great Britain.

73. Coxe, "A brief examination of Lord Sheffield's observations on the commerce of the United States of America," *American Museum* 9 (1791):121-126, 177-183, 217-226, 233-241, 289-295; 10 (1791):9-16.

74. As two recent studies illustrate—Merritt Roe Smith, *Harpers Ferry Armory and the New Technology* (Ithaca: Cornell University Press, 1977); and David A. Hounshell, *From the American System to Mass Production* (Baltimore: John Hopkins University Press, 1984)—there was no sudden conversion to the factory system in the 1790s or even a complete adaptation to machine tools somewhat later. Large-scale technological change still took time.

8 THE INVENTOR CELEBRATED

Americans after the War of Independence knew that they were not yet free of a lingering technological dependence on Britain, so they sought to push out on their own. Post-war enthusiasm for manufactures and agriculture brought a new and longer list of societies dedicated to technological improvement. The revival of these organizations paralleled an outpouring of scientific and technological interest bottled up during the war. Political independence seemed to release a kinetic energy. In a very real sense the American political experiment furthered the cause of American technology.

Americans exuberantly, albeit self-consciously, went about constructing their new nation. In the process they opened greater opportunities for "Yankee ingenuity."[1] Legislation supporting invention and technological innovation developed from a political freedom bringing an increased ability and inclination to act. Independence and mounting involvement in the Industrial Revolution boosted technology by prompting Americans to found new and refurbish old scientific organizations. Some Americans made a closer connection between science and technology, if not yet in the direct application of scientific principle to large-scale technological change, then in their celebration of science and technology as agents of progress. And those Americans began to think more seriously about invention. Independence and industrialization set the stage for the subsequent passage of patent legislation, first by the states, later by the federal government.

Four to five years before the war ended, patriots began talking about the place of science and technology in an independent America. David Ramsay made this subject a part of his lecture in Charleston on July 4, 1778. With independence, he declared, "The arts and sciences, which languished under the low prospects of subjection, will now raise their drooping heads and spread far and wide, til they have reached the remotest parts of the continent." Rallying his listeners to the standard of science and social progress, he asked, "May we not hope, as soon as this contest is ended, that the exalted spirits of our politicians and warriors will engage in the enlargement of public happiness, by cultivating the arts of peace, [and] promoting useful knowledge[?]"[2]

As if in response the American Philosophical Society, laid low by the vicissitudes of war, tentatively began collecting itself in 1780. On March 16, 1780, Timothy Matlack delivered "An Oration" before the society, the first such annual address since David Rittenhouse's in 1775.[3] He was followed by Owen Biddle in 1781, Thomas Bond in 1782, and Francis Hopkinson in 1783. Biddle gave a "historical sketch" of "improvements in useful knowledge" from ancient Greece to the Age of Discovery to show that invention, along with knowledge and understanding, had brought the rise of western civilization. He focused on science, but with the implication that science, technology, and invention went together. "All our valuable attainments depend on a diligent and close application in our pursuit after facts," averred Biddle, "and . . . one invention is linked in with and leads to many others which are remote and unforseen."[4] Thomas Bond lent an environmental twist to his address the following year in arguing that the brisk climate of America quickened the mind and nurtured a pre-disposition toward the liberal ideas leading to independence. Long-term independence, he added, was founded on science: "Point out the nation which has not Science, or that has abandoned it, and I will point out to you Savages and Slaves."[5] In 1783, the year of the Peace of Paris, Francis Hopkinson reflected that the philosophical society had begun with great promise but "was totally overcast by the dreary tempest of war. The still of philosophy could not be heard amidst the hostile din of arms." He called on his fellow society members to don once again the mantle of science and prove to a waiting world that the United States would be a showpiece of progress and light. With independence, he stressed, "It is high time that philosophy should again lift up her head."[6]

A Boston-based organization, the American Academy of Arts and Sciences, joined the rejuvenated American Philosophical Society in promoting science and technology. The academy linked political patronage with scientific investigation by inviting politicians with few or no scientific qualifications to become members. Although founded in 1780, the idea for the academy dated back to John Adams's days in Congress and his meetings with French scientists in 1779. The Massachusetts House of Representatives, General Court, Council, and governor all threw their weight behind the society's charter because the "Arts and Sciences are the Foundation and Support of Agriculture, Manufactures and Commerce." Advance in useful knowledge cast a good reflection on the government, they reasoned, and that advance was "most effectually cultivated, and diffused through a State, by forming and incorporating Men of Genius and Learning into Public Societies."[7]

Politicians probably backed the academy because of those utilitarian goals. Governor James Bowdoin was named as the organization's first president, a clear indication of the attempt to blend politics with science and technology in the American march toward social progress. Bowdoin may have been interested in broadening the fields of medicine, mathematics, astronomy, and natural history, but judging by the text of his presidential address, he was even more interested in what science and technology could do for agriculture, manufactures, and commerce. Technology would underwrite the new republic, be believed. Naturally Bowdoin emphasized that scientific societies were symbolic of social maturity and fitting tributes to national independence.[8] He and members of the academy as a whole nonetheless confessed that they owed an intellectual debt to Europe, whose scientific societies had generated "a spirit of discovery and improvement . . . among the ingenious."[9] This willing deference to Europe would not last many more years.

New patent legislation in several states came simultaneously with the wave of manufacturing, agricultural, and scientific societies washing across the land. Political independence gave the initial impetus to state patent legislation in the 1780s. Americans felt that they had to prove wrong those skeptical Europeans predicting their impending collapse. One way to do this was to reward ingenuity. Silas Deane, himself interested in securing patents for steam-powered grain and salt mills, was told in 1784 that there "is a disposition" among the "different Assemblies" to grant patents. Two Connecticut entrepreneurs had already received industrial privileges in the form of ten-year monopolies, a throwback to the type of patent common in the Colonial Era.[10] That same year South Carolina passed a patent act granting inventors fourteen-year monopolies for new and "useful machines," while Pennsylvania reversed its past policy of not awarding patents. In 1785 Arthur Donaldson finally received a patent from the Pennsylvania legislature for his dredge, and in 1788 George Wall patented his "Trigonometer" for surveying.[11] Maryland granted Robert Lemmon a patent for his spinning and carding machine in January 1787, and four months later awarded Oliver Evans two fourteen-year patents, one for his improved flour-milling machinery, the other for his "steam carriage."[12]

Inventors were heartened by the renewed surge of interest in scientific societies, the passage of patent legislation, and hints that they might receive more encouragement than heretofore. They may also have noted that patents came to relate more and more to the protection of intellectual property, to rewards for original genius, and less and less to industrial privileges.

As before the war, the American Philosophical Society led the way in promoting invention. It continued to solicit information on inventive activity from around the country, and members themselves tinkered with new devices. In 1786 the society accepted a stipend of two hundred guineas from a Portuguese member of the Royal Society of London, John Hyacinth de Magellan. Magellan's gift enabled it to offer ten guineas annually to any person making a contribution to natural philosophy or an improvement in navigation. Hope of winning the coveted Magellan premium brought forth a rash of inventors with steam engines, magnetic needles, and "spring blocks" for ships. Such incentives were all too rare for these ambitious souls.[13]

Equally as important, the general public took greater notice of inventions than in pre-war years. And there seemed to be more inventors and more inventions than ever before. A Connecticut mechanic had by 1780 designed a device to grind more wheat faster and use less water to drive the waterwheel and stones. Ezra Stiles claimed that "there are but four of these Mills in the world, all in Connecticut . . . the Inventor keeps it secret—but it ought to be divulged and he to receive an ample Rew[d] from the publick."[14] Some Connecticut residents sided with Stiles in viewing their state as a center of inventive bustle. After all, had not Nathan and Jotham Fenton of New Haven made a reflecting telescope and a microscope comparable to the finest instruments in Europe?[15] Not wanting to be outdone by someone in Connecticut or anywhere else, Philadelphia mechanic William Schultz let it be known that he alone had devised no less than a dozen inventions. Fellow Philadelphian George Turner, along with Richard Wells and one other, had begun experiments on a steam-powered cannon.[16]

The *Columbian Magazine,* in addition to promoting American manufactures, often carried pieces on inventors and on inventions ranging from agricultural machinery to a truss bridge designed to span the Schuylkill River.[17] Other magazines did likewise, though notices on more important inventions remained rare. Americans grew more conscious of the potential power of invention, but not suddenly. Through the 1780s they still may have perceived inventions as curiosities more than anything else.[18] The *American Museum,* the foremost journalistic proponent of economic nationalism, in six years of circulation did not carry one article on inventors or news of important inventions. Growth of manufactures, economic vitality, and inventive output were not yet seen as generically interconnected by most Americans. Entrepreneurs like Tench Coxe seeking to import British textile machinery understood the marketplace value of invention, but through the 1780s Coxe was part of a small if powerful minority. Moreover, eagerness to adopt inventions that had proven themselves in Britain did not simultaneously bring an eagerness to adopt new, unproven inventions made at home. The importation of British technology into the United States showed the profitability and utility of some inventions. With that example to build on, the public became more receptive to invention in general, though not until the

Revolutionary Era wound to a close. In the interim, inventors of even the most useful devices struggled for recognition.

Inventors working on internal improvements—bridges, canals, and steam transportation—were among the first to achieve public recognition. Bridge building provided an important outlet for inventive activity. Before the Revolution the only bridge project of note had been the Leffingwell Bridge, a modest truss and arch structure built in 1764 by John Bliss.[19] Compared with this, the 1780s experienced a bridge-building boom. In 1785 the Massachusetts legislature incorporated the Charles River Bridge Company. Samuel Sewall, designer, and Lemuel Cox, master builder, oversaw construction of the bridge, which took thirteen months. When completed, it stretched for a length of 1,503 feet. Its simple drawbridge could be raised by two men doubling as toll takers. Lemuel Cox added an innovation by soaking the pier timbers in oil, using oil to preserve wood against underwater rot and destructive worms. Cox later went on to a successful career building bridges in Ireland—an early turnabout in the transatlantic transit of technology.[20] The Charles River Bridge became a source of civic pride, and Cox became a local celebrity. Twenty thousand people witnessed the grand opening on June 17, 1786. The Malden and Essex Bridges, also built in Massachusetts in the late 1780s, similarly stirred the popular imagination.[21]

Lemuel Cox had found a receptive audience, a ready-made market for his bridge-building skills. Thomas Paine did not. Paine designed an iron bridge in 1786 for the Schuylkill River in Pennsylvania and lined up the moral support of Rittenhouse, Franklin, and other leading men of science. Yet the amount of iron and money Paine required was enormous, so enormous that the Pennsylvania Assembly could not afford to back him, private investors shied away, and ironmasters doubted their ability to deliver enough iron. Defeated, Paine went to England. In addition to securing a patent there, he received the acclaim of the French Royal Academy of Sciences. He eventually failed, but not before coming much closer to achieving what he had sought so futilely in Pennsylvania.[22]

The Massachusetts bridge projects of the same decade were funded because they were not as risky or costly. The bridges erected there were innovative, but not in the same way as Paine's single-arch iron structure. They had not challenged basic engineering theories and were built of wood, a common and inexpensive material. The deck was stacked against Paine. He proposed to do too much.

Oliver Evans, the most ingenious inventor in Revolutionary America, would have been justified in feeling that he had more cause for complaint than Paine. Born in 1755 in New Castle County, Delaware, the fifth son of a cordwainer-farmer, Evans demonstrated early an aptitude for mechanics. He first turned his thoughts to steam power in 1772 while serving

15. Oliver Evans, a genius among the ingenious. Smithsonian Institution Photo No. 32825-T.

as a wheelwright's apprentice. His ideas took definite shape after he read a treatise on the Newcomen engine. By 1773 he was seeking a practical way to make steam the motive power for carriages. Five years later he first contemplated using steam to propel boats. Though he lost interest in the latter project and years later chose not to contest either James Rumsey's or John Fitch's claim to originality, he continued to dwell on the application of steam to land travel. In the meantime, while finishing his apprenticeship during the war, Evans devised a technique to make wire from bar iron and designed a machine for cutting wire teeth for leather cards. That machine punched out five hundred teeth per minute, a rate surpassing Evans's most optimistic expectations.[23]

During this period Evans came up against three obstacles that he complained blocked him until the turn of the century: public scorn for his genius, inadequate patent protection, and widespread antipathy to new mechanical devices. Evans found this last obstacle in particular difficult to clear away. All too many thought of Evans as a harmless visionary, an eccentric who rambled on about steam engines and a transportation revolution.[24] He had been ridiculed by friends and family alike when he disclosed his plans for a wire-cutting machine. He overcame their derision by proving the practicality of his scheme but then ran into another snag. His application to the Pennsylvania Assembly in 1778 for a $500 loan to build a mill using his machinery died in committee. Meanwhile word of Evans's innovation spread quickly, and since he was not shielded by a patent, his machinery design was copied, and he made little money. The outcome was consistent with the rest of Evans's early career.

Shrugging off his disappointments, Evans progressed from his card-making machine to design a fully automated grain mill. In 1783 he scraped together enough cash to build a mill employing his new machinery on the Red Clay Creek in Delaware, the breadbasket of the Middle states. His milling refinements, most notably an "elevator" (an endless belt carrying grain from one level to the next), a "conveyor" (essentially an Archimedean screw moving the grain horizontally), and a "hopper-boy" (for raking the grain) dispensed with the labor of four or five men. While other versions of the elevator and conveyor had been used earlier by some millers, most notably by Evans's close neighbors the Ellicott brothers, Evans combined methods and machinery in a uniquely efficient way. One worker could run his entire water-powered mill. The grain, untouched by human hands once it entered the machinery, was ground more methodically and produced better flour.

Evans had two goals. First, he wanted to protect himself by obtaining patents. In this he was comparatively successful. He secured them in several states, with Pennsylvania and Maryland granting him fourteen-year monopolies and Delaware a fifteen-year monopoly.[25] The second goal of persuading millers to adopt his machines proved more elusive.

Despite the proven quality of his inventions, Evans struggled for the better part of ten years to sell his flour-milling techniques. His brother Joseph toured Pennsylvania, Delaware, Maryland, and Virginia, offering to install the machinery free to any miller willing to introduce the technique to others. The Evans brothers had no more success than John Hobday had twenty years earlier in trying to interest Virginians in a new thresher.[26] Even the Brandywine millers in their own neighborhood resisted change and the introduction of Evans machinery until after 1791. Until then Evans only converted a handful of them (and George Washington, who worked Evans machinery into his Dogue River mill). Evans, ironically, experienced his greatest disappointments only a few years before his friends anxiously kept track of the introduction of Arkwright-type machinery in Pennsylvania and New Jersey cotton mills.

A dejected Evans sold his automated mill and machinery and moved to Philadelphia, where he hoped to enlist congressional support. After twenty years of trying to make a living as an inventor, he had little to show. His mill machinery was patented in several states but was far from being universally employed. His steam carriage scheme was hardly taken seriously, and only Maryland saw fit to grant him a patent for it. Other state legislatures did not deem it worthy of their consideration. Evans persevered, perhaps because he felt that "if Providence has endowed me with a Genius Capable of Invention Probably I may render My Country Great Service in this than any other line I can engage in."[27] Eventually he made a comfortable living, thanks to his pioneering work in high-pressure steam engines and tardy adoption of his mill machinery. In 1805 he finally demonstrated his steam-driven "Orukter Amphibolos" for Philadelphians who were as astonished and unbelieving as the Parisians who gawked at Nicholas Cugnot's steam carriage in 1769.

After the War of Independence more and more Americans saw the connection between invention and improved technologies, though not fast enough to suit inventors. Oliver Evans and his colleagues remained less than satisfied with their lot. Part of their dissatisfaction might be dismissed as peevish hypersensitivity, for inventors in the Revolutionary Era often had an exaggerated sense of self. Evans was never happy with the way that he or his inventions were treated. Impatient to succeed, he and other inventors sometimes pushed projects beyond reasonable limits.

Political independence by itself had not brought (indeed could not bring) a sudden change in public attitudes toward invention. Enthusiasm for material progress and the passage of patent legislation was juxtaposed with a lingering public perception of invention as the source of curiosities and nothing more. Even if Evans's automated mill was of tremendous technological importance, then and there, most millers did not rush to buy it. Their reluctance to retool could have resulted from a shortage of

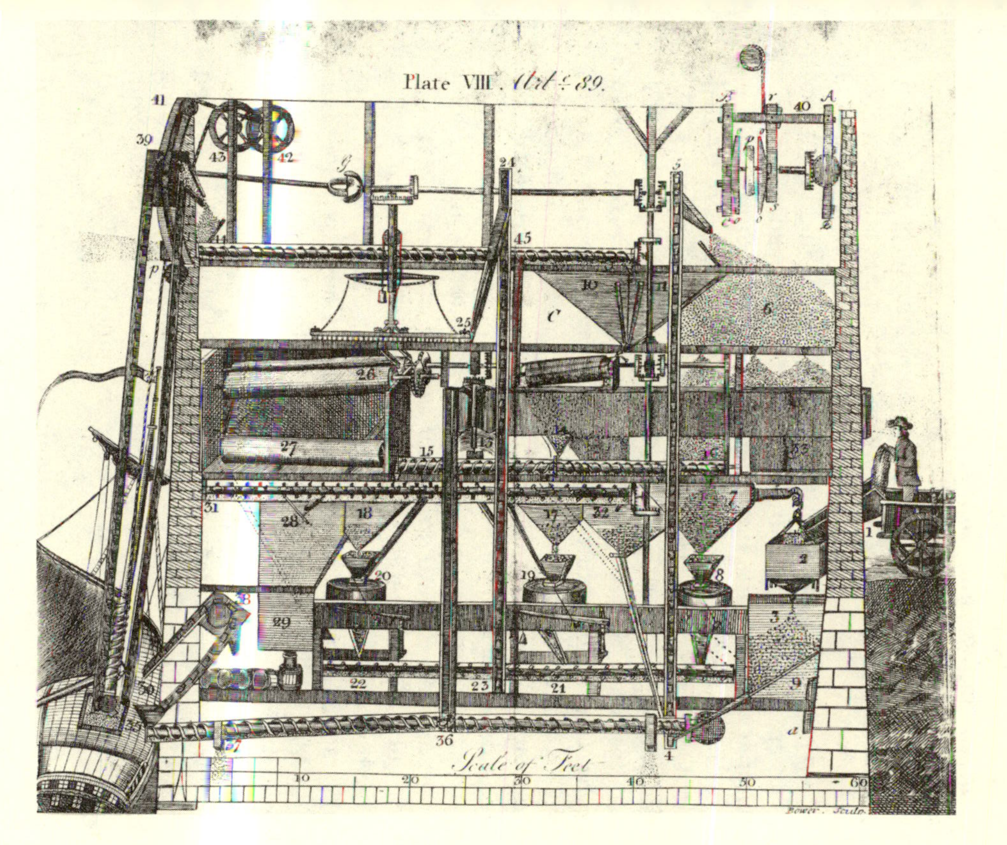

16. Oliver Evans's automated flour mill, from *The Young Mill-wright and Miller's Guide*, 3rd ed. (Philadelphia: M. Carey and Son, 1818). As Evans noted, this drawing was "not meant to shew the plan of a mill; but merely the application and use of the patented machines"—notably the elevator, the conveyor, and the hopper-boy.

investment capital, technological apathy, or any number of other reasons. They can hardly be dismissed as a group opposed to technological change because many were innovative in their own right. Innovative or not, to them the Evans method may have seemed too great an investment in change.

Evans soon learned that his troubles did not end with reluctant millers. Like other inventors, he found that state patents, while a step in the right direction, if anything fomented rather than checked damaging competition. States awarded patents for different lengths of time. Pennsylvania granted them for as many as twenty-one years and as few as five.[28] In addition, states had not agreed on what was patentable and had not set up procedures to test originality. The state patent system of the 1780s was an albatross, a reminder that Americans had only begun to approach invention as part of a larger technological whole.

The national government as it operated under the Articles of Confederation added another political irritant to the list of inventors' woes. The Articles said nothing about patents or government support of technology, so Congress under the Articles reacted to inventors inconsistently. It held out the carrot to some, turned others away abruptly, and annoyed them all by its erratic behavior. Inventors hoping that independence would bring with it a central government ready and able to assist them were soon disabused of that notion.

Evans's chances were also hurt by crank inventors. Quixotic John Macpherson survived the war to pester leading statesmen with his schemes. In July 1787 he notified Congress that he had discovered a foolproof method of measuring longitude by using a magnet and "dipping needle" rather than by heavenly observation. Congress ignored his petition.[29] In May 1788 he again made contact, this time to report he "hath Invented a Fort, that one Thousand men can with ease, and with safety, defend against all other men on earth." This plan too was pushed aside.[30] Penniless, rebuffed, and near death in 1791, Macpherson begged Henry Knox to persuade Congress to grant him a loan. Never giving up, he assured the Secretary of War he could "pay what is lent with Interest, soon after my Inventions were examined, and my Patents granted."[31]

Macpherson was one of several inventors to bother Congress with impractical schemes.[32] Unfortunately, he may have helped to give inventors a bad name. Although he did not exclude himself from the ranks of "semi-lunatic Projectors," Francis Hopkinson made some sniping remarks about inventors in general. He indiscriminately placed steamboat pioneer John Fitch in the same category as men like Macpherson. David Rittenhouse—himself an inventor—was less sweeping in his criticism, yet he did complain that "we have an abundance of projectors and pretenders to new Discoveries," some of whose ideas were "ridiculous enough."[33] The public could hardly be expected to affectionately embrace new devices when inventors themselves could not agree on who was a legitimate inventor and what was a legitimate invention.

Congress found that it was easier not to get involved. In October 1781, the month Cornwallis surrendered at Yorktown, James Hopkins sent a memorial to Congress requesting a premium for his newly invented card-cutting machine. His memorial was referred to a select committee, never to be heard from again.[34] Trenton steel manufacturer and inventor Stacy Potts applied to Congress in May 1783. He reminded the nation's leaders that even though political independence had been achieved it still remained for them to support improvement of "useful manufactures" to protect that independence. Potts asked for neither an "exclusive Priviledge" nor a premium, both of which "are inconsistent with our freedom." Instead, he wanted Congress to examine his steel and, if it found the metal to be as good as he claimed, to recommend its use. Potts sent along affidavits certifying that he made a high-quality product. Congress appointed a committee to review Potts's petition, but that committee ruled that to support one manufacturer to the exclusion of others, even if no money changed hands, was discriminatory. Questions of bounties, special privileges, or recognition were best left to the states.[35]

After the war Christopher Colles devised several schemes, but he too came up against what he felt was a brick wall in Congress. In 1783 he submitted a plan to open the Pennsylvania and Virginia backcountry to commerce by improving navigation on the Ohio River. How he determined to do this, he did not state clearly; he did hint that a canal detouring a three-mile section of treacherous rapids had to be dug. In return for his work as chief engineer, he asked Congress for a parcel of land along the river. Congress did not encourage him, and neither did George Washington. Colles visited Washington, thinking the retired general could interest Virginia legislators in his plan. Colles did no better here, for little did he know that Washington and other leading Virginians entertained western plans of their own, but for the Potomac, not the Ohio River.[36]

Colles scrapped his Ohio River plan and came up with a far more ambitious proposal. He tried to persuade New York investors to finance the clearing of the Mohawk River and by a circuitous route link it with Lake Ontario. His goal was to open an uninterrupted channel from the Great Lakes to the Hudson River, building canals where necessary, to lure settlers west. His scheme to finance the undertaking sounded deceptively simple. The New York legislature would grant investors 250,000 acres in exchange for $13,000. The investors would eventually get their money back by selling the land to incoming settlers. Colles, the chief engineer, carefully set aside substantial properties for himself.

The sheer magnitude of the project, which Colles unrealistically guaranteed would take no more than three years and five hundred laborers, frightened away potential investors.[37] An internal improvements program on this scale surpassed the financial ability and comprehension of Colles's contemporaries. Colles himself may not have fully understood the undertaking he advocated. His vision dissolved, he devoted his energies to a survey of American roads from Albany to Williamsburg, a useful if much less satisfying job to Colles personally. The year 1789 found him in New York, short of money, and fruitlessly lobbying Congress for a patent on one of his many inventions.[38]

Colles waited for congressional action with a clutch of inventors who virtually inundated Congress with petitions in the spring and summer of 1789. Joining Colles were Englehart Cruse, who had an "improved steam engine, which he has invented, for raising water for the purposes of mills" and factories; Leonard Harbaugh, the designer of a new thresher, a new reaper, and a dredge; and no fewer than six Philadelphians: John Stoebel and Alexander Lewis, promoters of two new modes of upriver travel; Samuel Briggs, the inventor of a nail-making machine; Arthur Greer and John Churchman, discoverers of purportedly unique ways to measure longitude; and, pathetically, John Macpherson, who claimed to have improved Franklin's lightning rod.[39]

Colles, Cruse, and the rest pushed for patents and special privileges because by 1789 the national government had changed. The Articles of Confederation had been shelved and replaced by the Constitution. The Constitution strengthened the central government, and, of particular interest to inventors, certain key phrases in the document reflected its authors' commitment to science and technology. Article 1, Section 8, stated that Congress was empowered to "promote the Progress of Science and useful Arts, by securing for limited Times to Authors and Inventors the exclusive Right to their respective Writings and Discoveries." This passage implicitly paved the way for federal copyright and patent laws. It grew out of Charles Pinckney's proposal at the Constitutional Convention that Congress be authorized to "grant patents for useful Inventions" and James Madison's recommendation for "premiums & provisions" to advance "useful knowledge and discoveries."[40]

The first federal patent statute, passed on April 10, 1790, gave form to the promises made in the Constitution. It reflected congressional concern that the federal government do something to protect the rights of inventors and define the limits of intellectual property, without passing out rewards or contracts.

Inclusion of a patent clause in the Constitution derived from more than just a theoretical objection to the lack of a similar provision in the Articles of Confederation. Members of the Constitutional Convention realized that leaving patent legislation to the states was an inefficient and unfair mode of operation. States drew up their own criteria for patents, determined their length and renewability, and awarded them as a form of special legislation. In states like Pennsylvania and Virginia, the length and details of a patent varied from one inventor to the next. With more and more inventors seeking patent protection, the founders probably foresaw the unavoidable overlap if such a system were kept intact and the acrimonious litigation that would result.

The reality of this was brought home to them in the John Fitch-James Rumsey steamboat controversy, an inventors' war lasting seven years. The steamboat controversy involved the federal government, numerous state legislatures, and most of the country's leading scientists. Fitch and Rumsey's heated struggle riveted public attention. Their feud "tested the intention of national leaders of relying primarily upon market forces and the patent system to encourage the mechanization" they were convinced was essential to national prosperity.[41]

The seeds of the Fitch-Rumsey controversy were sown in 1783. In July Rumsey's friend James McMechen of Virginia requested Congress to grant him one thousand acres from the public domain north and west of the Ohio River. He wanted the grant as a reward for his "mechanically" powered boat, which was designed to travel against river currents without the aid of a sail. McMechen was not specific about the motive power to be employed, but a congressional committee was sufficiently intrigued to recommend that he be granted a premium—money, not land—if his boat performed as promised. The committee added apologetically that "tho' Congress must have the greatest desire to encourage every Mechanic Art, which may tend to the promotion of Trade and Cultivation of this great Empire," a grant of land would be improper.[42] McMechen had every reason to be optimistic, considering how precipitately Congress had disposed of Colles, Potts, and Hopkins. Members of Congress seemed to be more interested in invention than they had been a few years before.

McMechen's petition was still pending a year later when Congress heard from a different mechanical boat inventor. The new communicant was Rumsey himself, then engineer of the Potowmack Canal Company, George Washington's ill-fated western navigation–internal improvements project. McMechen delivered the petition to Congress for Rumsey. Apparently he and Rumsey had collaborated all along. The ambiguity of Rumsey's wording in this petition set the stage for the steamboat controversy because he later contested that he had implied the use of steam power.

Congress went along with Rumsey's request for thirty thousand acres, perhaps because he made it clear that he expected nothing if he failed. Either Congress was certain he would indeed fail, or it reversed its position on the awarding of western lands to inventors. Thus, if he built a "mechanical boat" of ten tons' burden that traveled up the Ohio River fifty miles a day for six days, and with a maximum crew of three, he would get his choice of thirty thousand acres in the public domain.[43] By the time Congress deliberated and decided, a fast-working Rumsey had already secured patents in Pennsylvania and Virginia. The Virginia patent, good for ten years, was the first to be granted in that state after the war.[44] Rumsey innocently—or cagily— left the wording of his petitions to the two states ambiguous; steam is nowhere mentioned and is nowhere expressly ruled out.

John Fitch entered the lists in 1785. He first considered using steam for land carriages in April of that year then "gave it over as impracticable" and turned his attention exclusively to steam-propelled boats.[45] He applied to Congress in August 1785 and appended letters of support from three prominent academicians. He asked Congress to examine his proposal and "judge whether it deserves Encouragement." Compared to the treatment accorded Rumsey, Congress dealt curtly with Fitch. His petition was referred to a special committee, but the committee was dismissed, and Fitch's papers were returned to him within a few days.[46] Deeply offended, Fitch offered details of his project to the Spanish Minister residing in New York. This was no doubt a ploy to shock Congress and show the committeemen who reviewed his petition that they were "but Ignorant Boys" and "Blockheads," not philosophers or mechanical geniuses.[47] Fitch also communicated news of his steamboat idea to the American Philosophical Society.[48]

Fitch petitioned Congress a second time in March 1786. Less deferring in this petition, he requested a patent granting him "the exclusive Priviledge of constructing Boats impelled by the Force of Steam, and the advantages arising from that discovery, on all the Waters now belonging to the United States." Fitch wanted to monopolize all forms of steam travel on water, with a patent extending him that monopoly indefinitely. Congress, having no explicit authority to grant a patent, held the petition three weeks and then filed it away.[49] It avoided taking a clear stand on the patent issue in what was fast becoming a controversial case, but not in a way calculated to appease Fitch.

Fitch had gone on the stump after Congress turned him down the first time, taking his case to the individual states. While touring Virginia and Maryland, he heard about Rumsey. In 1784 Rumsey had made a mechanical pole boat driven by the river current; that much was incontestable. What became an issue was Rumsey's assertion that he had conceived of using steam power at about the same time—at any rate, months before John Fitch. The rival claimants, now openly at odds, scurried about gathering support from politicians and scientists. They also fought each other for state patents.

A neurotic Fitch felt certain there was a conspiracy against him. He charged George Washington and Governor Thomas Johnson of Maryland with deceitfully giving credence to Rumsey's unsubstantiated claims. He soon after added Benjamin Franklin, Francis Hopkinson, and Arthur Donaldson to his enemies list. Fitch maintained that he had explained his ideas to Franklin in 1785 and Donaldson in early 1786 and that Donaldson, with Franklin's and Hopkinson's complicity, had used that information to design a "jet boat" powered by steam.[50]

The Pennsylvania Assembly found for Fitch, and did so without imputing guilt to any of those Fitch had accused of doing wrong. With this issue resolved, Fitch returned to the bigger contest with Rumsey. From 1786 to 1789 he and Rumsey marshaled their forces by organizing "companies," did everything in their power to discredit the opposition, and attempted to build working steamboat models. Fitch jumped off to an early lead by securing patents in New Jersey, New York, Delaware, Pennsylvania, and Virginia for all forms of steam navigation.

By granting Fitch a patent, Virginia and Pennsylvania showed they rejected Rumsey's contention that he had implied using steam power back in 1784, when those states granted him monopolies on his "mechanical boat." Yet Rumsey did not lose out completely. He kept his 1784 Virginia patent for a "mechanical boat." Fitch's 1787 Virginia patent for steamboats, though granted for fourteen years, was issued with a rider stipulating that he had to have a steamboat operating on a Virginia River within three years or his patent would become void.[51] Rumsey had no such conditional clause in his ten-year patent. Like Fitch, he nevertheless felt slighted because Virginia issued patents as special legislation rather than under a uniform code. His 1784 patent was almost worthless because his pole boat did not work as well as he had hoped.

When looking for a skilled craftsman to make his steam boiler, Fitch thought about approaching Christopher Colles before settling on Henry Voight of Philadelphia.[52] He made several full-scale models but was frustrated by technical problems, primarily leaking boilers and insufficient speed for the Delaware River, where he conducted his tests. By 1788 he admitted that £1600 had been "expended to but little purpose."[53] Fitch had been encouraged, if only momentarily, by a successful trial run on the Delaware in August 1787. David Rittenhouse, Richard Wells, and a cluster of politicians—some of whom were sitting in the Philadelphia Convention—witnessed the event and urged Fitch to press on. He applied to Congress a third time in January 1788 and at last produced some results. Congress allocated western lands of his choice as a reward for his trial run and promised more acreage if he built a steamboat that operated on a regular schedule.[54] The difference between its handling of Rumsey in 1784 and Fitch in 1785 and 1788 is that by the latter date Fitch had gone from theory to practice. Bearing this in mind, Congress's treatment of Fitch and Rumsey was, technically at least, consistent, which is more than can be said of some states. Nonetheless, both inventors

17. John Fitch's steamboat, with the basic paddle design used in the 1787 test on the Delaware River. From the *Columbian Magazine,* December 1786; courtesy of the William L. Clements Library, University of Michigan.

resented the federal government as established under the Articles of Confederation, regardless of the land grants, because it did nothing to resolve their dispute and bring a semblance of order to the patent system. Fitch and Rumsey barnstormed the country, exhausted their resources, yet settled nothing in the end. Their resentment was therefore as deep as that of inventors like Christopher Colles who had received no congressional encouragement at all.

Deed in hand, Fitch enjoyed a measure of success to go along with his larger dose of frustration and failure. After a delayed start, Rumsey meanwhile labored mightily from 1786 on in a losing battle to gain acceptance as the "true" steamboat inventor. Maryland granted him a patent early in the contest, but Fitch had beaten him to the key states. Rumsey staged a trial run in December 1787 with a boat moved by steam and like Fitch lobbied wherever he could.[55] He trekked to Philadelphia in 1788, where he met with representatives of Fitch's company. They reached no compromise. Fitch refused to share credit for the discovery or modify his aim to monopolize all forms of steam-powered water transportation.

To further widen the breach, both men used their steamboat companies to rally support. They split apart the American scientific community. Rumsey staged a coup by getting Franklin to join his "Rumseian" society.[56] While in Philadelphia, Rumsey also disclosed his intended application of steam to a variety of devices, hoping thus to discredit the monomaniacal Fitch.[57] The Fitch-Rumsey fight spread to the floor of the American Philosophical Society. A special committee ruled sensibly that the ideas of both men had merit, "but the best application must be [a] matter of experiment."[58] The decision was wise not only because the question of originality could not be resolved, but also because many society members were partisans of one side or the other. Nonetheless, since Rumsey had earlier been invited to join and he had not, Fitch viewed the noncommittal decision of the society with pained displeasure.

Rumsey had already initiated a pamphlet war with Fitch to prove his "prior right." Fitch of course denied that Rumsey had a case and published a rebuttal with documents to counter his opponent's evidence.[59] Refusing to let Fitch have the last word, Rumsey's brother-in-law and attorney, Joseph Barnes, published his own pamphlet wholeheartedly backing Rumsey. He argued with some justice that Fitch had brought his troubles on himself by insisting on a monopoly for all forms of steamboat propulsion. Rumsey, he pointed out, used steam as a direct motive force for his "rocket boat" while Fitch used steam to drive oars and paddles. Since the two inventors employed steam in different ways, they should have been able to reconcile their differences and cease their patent fight. This would in fact be the position taken by the federal patent office eventually set up in 1790. Unfortunately, Fitch (and then Rumsey, in response to Fitch's intransigence) demanded credit for original conception and application. Barnes's mixture of polemic and sound reasoning did nothing to clear the air.[60] With no federal patent office to end the dispute, Fitch and Rumsey continued to play one state patent against the other, like poker chips.

Neither man, as it turned out, enjoyed long-term success. Rumsey left for England in 1788, hoping to be better received there. He had gotten no land from Congress, and all his state patents eventually expired because he failed to construct a mechanical boat capable of carrying commerce. In England he negotiated with Boulton and Watt to build a steamboat, only to find that he had not escaped his two biggest problems outside Fitch: insufficient money and a dearth of workmen able to make a sound steam boiler. The latter was "the case I believe with all new inventions," sighed Rumsey, and his spirits sank.[61] Boulton and Watt, for their part, respected Rumsey the inventor but became disenchanted with Rumsey the businessman and severed their ties with him. They were well aware of the complexities of technological innovation and the need for caution. Their erstwhile American partner lacked their patience and could not wait. Rumsey died in England in 1792, embittered by his experiences on two continents.

Fitch ultimately met a similar fate. He ran a steamboat on the Delaware for a few successful months of passenger service in 1790, but his company went bankrupt and folded. His boat ended its short career as flotsam and jetsam. After all his troubles, Fitch was finally brought down by lack of sustained public interest in his steamboat ferry service. Hector St. John Crèvecoeur's prediction that he would suffer "a great deal of trouble and anxiety" had been fulfilled.[62] Crushed, physically and financially broken, Fitch wandered over to Europe and back and in 1798 committed suicide while in a self-imposed Kentucky exile.

While neither Fitch nor Rumsey prospered at their work, they could have taken some consolation in the knowledge that they had a part in bringing about the federal patent system. Both had been granted federal patents in 1791. Fitch received one for his "improved boiler of a steam engine," and Rumsey received six for his various adaptations of steam power.[63]

The Fitch-Rumsey controversy flared up during the Confederation Era and still smoldered when Congress assembled in 1789 and drew up legislation leading to the Patent Act of 1790. Fitch, in fact, had been among the inventors besieging Congress in the spring of 1789.[64] The steamboat dispute brought into focus the difficulties confronting any patent system. Rumsey confided to Thomas Jefferson his fear that he and Fitch had done much to sour Americans on the patent system as it stood before 1790. If he was right, it was just as well.[65]

Like most leading Americans, Jefferson had not remained aloof from the Fitch-Rumsey controversy. He sided with Rumsey. In fact, he respected Rumsey and Thomas Paine (because of his iron bridge) as authors of the two most important inventions in the young republic. In his estimation Rumsey was "the most original and the greatest mechanical genius" he had ever seen.[66] Perhaps Rumsey's troubles helped Jefferson rethink his own ideas about patents, for as late as 1788 Jefferson was averse to them as a form of monopoly injurious to the public. A year later he modified his stand to make an exception for patents to inventors granted for limited times. He concluded that a patent to an inventor did not steal something from the public; rather, it protected the right of an individual to enjoy the rewards of something the public would not have without him. In reaching this conclusion, Jefferson merely adopted, consciously or not, the viewpoint of Sir Edward Coke, who as chief justice under Charles I had made the same distinction between monopolies and patents in a legal discourse of 1628.[67]

Jefferson may not have been as enthusiastic about manufacturing as Hamilton and Coxe, yet he was every bit as much an advocate of technological transfer and a supporter of invention. Like Paine before the war, Jefferson embraced invention with nationalistic pride. He resented Britons who felt superior to Americans and was anxious to prove that Americans were in no way inferior. He was nevertheless torn. He wanted Americans to develop their own inventive skills and technological genius, but he also realized that Europe had much to offer his countrymen technologically. He was fascinated by the farm machinery he had seen in France and the gunmaking techniques of Blanc, and he hoped Americans would learn from them. He did what he could to learn for himself, both while abroad and back home in Virginia.[68] Monticello was littered with Jefferson's devices, and as Secretary of State Jefferson would later review carefully all of the patent applications and models that crossed his desk. In sum, Jefferson immersed himself in inventive activity and followed eagerly the experiments of others.

George Washington, another active participant in the Fitch-Rumsey dispute, had urged passage of the federal patent act in his first annual address to Congress on January 8, 1790. Washington was not one to let misguided nationalism blind him to the need for Americans to seek new ideas from abroad, to keep alive the transit of technology. "I cannot forbear intimating to you the expediency of giving effectual encouragement as well to the introduction of new and useful inventions from abroad," he counseled, "as to the exertion of skill and genius at home."[69] He joined Jefferson and concerned citizens like Noah Webster in advocating a federal patent code. Tench Coxe was another of those concerned citizens. Coxe wanted federal patents to protect the machinery designed by British artisans he had recruited. He wanted Congress to reserve one million acres as a fund to reward "the introduction of Machinery, inventions, arts and other things of that nature," whether "by Natives or foreigners."[70] As Webster noted and Coxe no doubt concurred, "the want of some regulation for this purpose may be numbered among the defects of the American Government."[71]

The Fitch-Rumsey controversy did not create this feeling, but it did outline the state patent system's inadequacies in bold strokes. It impressed on all parties the need for something better. It pushed the national government to take charge of the patent system faster than it would have otherwise. And as Brooke Hindle has concluded, "More notable than the fumbling manner in which America struggled toward the realization of the steamboat was the still rising enthusiasm for mechanization wherever it might conceivably work."[72] That enthusiasm, first sparked by the importation of textiles machinery, continued apace.

The Patent Act of 1790 launched a system more efficient than what it replaced. Under this act, patent applications were to be reviewed by a three-man board consisting of the secretaries of state and war, and the attorney general. These "Commissioners for the Promotion of Useful Arts" could issue patents for a fixed period of fourteen years, with no provision for extension. All applicants had to accompany their request with a written description and a model where feasible. Patentees were protected against infringement by the promise of a court trial, the guilty party to pay legal fees. Any patent could be revoked up to one year after the issue date if the board learned that it had been obtained fraudulently. Filing a claim could cost as little as $3.80.[73]

Federal patents superseded old state patents, which soon after became void, and brought long-awaited regularity to the system. Apparently as a result of the Fitch-Rumsey dispute, the states accepted the new patent code with some relief. Shortly after the 1790 act took effect, board member Thomas Jefferson noted that it gave "a spring to invention beyond my conception."[74] Inventors eagerly sought protection under the new system. In addition to the Fitch and Rumsey patents on steam power, George Parkinson patented a flax-spinning machine, and William Pollard patented an improved Arkwright frame. Thirty-three patents in all were taken out in 1791 alone. A dozen were for steam power, from boats to mills, and Oliver Evans at long last secured the legal protection for his milling innovations that he had been seeking for nearly a decade.

The superiority of federal legislation notwithstanding, some problems remained. Weaknesses in the 1790 act surfaced almost immediately. The three-man board was too busy with other business to meet regularly. Inventors objected to their not having an option to appeal a board decision. John Fitch charged the commissioners with twisting their interpretation of what constituted an invention to favor James Rumsey.[75] Rumsey, by then in England, was himself far from pleased with the act. He lamented that "the law of Congress respecting patent rights almost amounts to an Exclusion of my ever returning again to my Country."[76] Richard Wells, long-time inventor and loyal friend to Fitch throughout the steamboat controversy, objected to certain provisions for cornering privileges on imported inventions. Oliver Evans headed a group of inventors protesting the fourteen-year limit as so restrictive it "would sink them into despair."[77] Joseph Barnes filed the most impassioned protest in 1792, decrying the patent system as it existed and the alterations written into a new act pending in Congress. Barnes wanted to see an entirely new arrangement, with a special officer appointed to oversee patent disbursements. He grumbled that, as it was, the United States "tho' first in freedom" had taken too long to "establish an effectual system, for promoting science and the useful arts." Britain owed much of "its eminence in manufactures" and "the greater part of its wealth" to its patent system. That the United States, a republic among monarchies, should do less for its citizens was a national disgrace, he reproved. Invention, the source of material strength, the catalyst of progress, had to be jealously guarded.[78]

Congress responded with a new patent act in 1793, although Barnes thought that even this new act was inadequate. Patents were put firmly under the jursidiction of the State Department instead of a three-man commission, thus streamlining the bureaucracy. Patent applications were to include written descriptions and a model where feasible, and patents were granted for fourteen years, again with no provision for extension. But patentees benefited because disputed cases could now be appealed to a three-man arbitration board, with one representative chosen by each party in the dispute and the third appointed by the secretary of state. The board's decision would be final. Patent officers benefited under the new act because they could repeal patents obtained falsely any time up to three years after they had granted them. Patent violators would be penalized three times the cost of a legal license issued by patent holders if caught. As before, the guilty party paid all court fees derived from litigation, and the cost of filing was raised to a flat rate of $30.[79]

The new act did not resolve all problems and completely satisfy inventors. Joseph Barnes had lambasted it before it became law and continued to criticize it after it took effect. Oliver Evans claimed to be as dissatisfied with this new law as he had been with the old. He melodramatically threatened to tie his inventions and plans in a bundle and file them away "because a patent in this country is not yet worth the expense of obtaining it."[80] Melodramatic or not, Evans did have a point. Enforcing the new law was difficult and Evans, like other inventors, became involved in bitter disputes with patent infringers. Although the 1793 act required written descriptions and models and even an oath swearing sole authorship, it did not establish procedures to determine proof of novelty. Applications were not closely investigated and were usually accepted at face value. Patent officers could grant patents to several different inventors for essentially the same invention, a confusing possibility not clarified until the reforms of 1836. Many inventors also resented their inability to secure an extension to the fourteen-year limit, and some criticized the $30 fee as exorbitant.

Yet no patent act could answer all the perceived needs of Oliver Evans and other inventors, for their biggest worries could not be removed by legislation. Inventors felt that as a group they were too little respected, their ideas too little appreciated by the general public. No amount of legislation could change that. Inventors had a hard time understanding that, since they projected new schemes, they had to endure the seeming unresponsiveness of a public that trailed behind them.

For some inventors of the Revolutionary generation, federal patent legislation came too late. They were either too advanced in years or too disillusioned by experience to think much of it. A federal patent system, the outgrowth of American society's budding commitment to technological progress, did not bring with it bliss for all inventors. In the end public indifference defeated John Fitch, patent protection or no. Inventors might well have heeded Crèvecoeur's advice to a despondent Fitch that "such . . . is the fate of all inventors, such is the Price at which Nature sells talents of genius, like a woman in child birth you must expects pangs and panics."[81]

Passage of federal patent statutes did not endow inventors with sudden wealth. Oliver Evans and others like him spent years and fortunes in endless litigation. Still, the extent of their legal troubles should not be exaggerated. Evans, his many protestations to the contrary notwithstanding, did not end his days poverty-stricken and forgotten. He was a member of the first generation of American inventors to become famous for being inventors. He, like Fitch and Rumsey, had helped move inventors out of isolation. They lost their anonymity and became recognizable participants in the American quest for technological independence.

In the larger scheme of things, the patenting career of an individual inventor like Evans is less important than the motives behind patent legislation. Here the Americans patterned themselves after the British, though not closely enough to suit Oliver Evans or Joseph Barnes. The French government bestowed rewards and privileges on inventors far more lavishly than did the British, yet the Industrial Revolution did not spread to France until after it was well under way across the channel. Invention and patent granting in Britain had been brought more fully into the mainstream of technological change because of basic differences that distinguished the two societies, from labor and business practices to religious and community customs. It was the cumulative effect of these individual components of society that charted the course taken by Britain, no less than by France and the infant United States. In the United States, patents became an integral part of a larger technological process, a reflection of the national commitment to material progress. Alexander Hamilton had shown the way in promoting the cause of invention, and for that very reason. In his *Report on Manufactures,* he predicted that inventors would cure the economic ills of the new nation by introducing new machines and making possible new industries. Inventors were to be drawn into the very practical business of industrialization.

Of course not every tinkerer became an industrial innovator after 1790. And it was not until much later, near the end of the nineteenth century, that inventors would become fully integrated into the economic and technological growth of industrial America. Still, by 1790 those possessing inventive genius had a clearer frame of reference and a greater chance of success than in the Colonial Era. American society had become more technologically inclined. The inventor gained more and more attention until eventually, with Thomas Edison, the inventor came to symbolize to Americans the superior genius of their society.

Inventors of the Revolutionary Era could have predicted as much. John Fitch did not make his fortune, but he did foresee the future veneration of invention in the new nation. He wrote, with perhaps grim satisfaction, that inventors "only have raised us above savage Barbarity."[82] Other inventors felt the same. "It is by the improvement of the arts and sciences, as applicable to agriculture, commerce and manufactures," echoed Oliver Evans, that national wealth and power would be secured.[83] Their predictions were among the most bittersweet prophecies of their generation.

Joel Barlow, who had not suffered the frustrations of Evans and Fitch, waxed poetic on the theme of invention in the new nation:

> While rising clouds, with genius unconfined,
> Through deep inventions lead the astonish'd mind,
> Wide o'er the world their name unrivall'd raise,
> And bind their temples with immortal bays.[84]

Barlow's ode depicted political independence as the first step in the liberation of America. A chosen people, once held in bondage by Britain, would now leap forward in a creative burst. Ezra Stiles and the Marquis de Chastellux had shared a similar vision, even before the new nation was officially independent. In 1783, several months before the Peace of Paris ending the War of Independence was ratified, Stiles wrote a passionate address on the future of America. Americans, he emphasized, were virtuous people in a plentiful land. Now free to determine their own destiny, they needed only to live up to their promise of future greatness. With republican political institutions, a commitment to civil and religious liberty, and the pursuit of economic opportunity, they could become a truly righteous people of great wealth. And part of their greatness would be based on the promotion of the "mechanical arts"—technology. "American inventions and discoveries," he proclaimed, would help to insure national prosperity. Stiles boasted of American inventive skills and urged that they be applied to manufactures, agriculture, and commerce.

Just before setting sail for France, Chastellux also predicted a bright future for the new nation. Writing in the cabin of a French frigate anchored in the Chesapeake, in sight of the land he was about to leave, he wrote an essay on "The Progress of the Arts and Sciences in America." Having observed the practicality and industriousness of his hosts, he concluded that all of the sciences, mechanics included, would prosper in America. Like Stiles, Chastellux believed that the American environment and the American people were perfectly suited to advances in every field.[85]

And thus Chastellux, Stiles, and Barlow tied the fortunes of the new nation to the progress of science and technology as well as to abstract ideas of liberty and equality. Technology could be used to foster republicanism, they decided. While Chastellux and Stiles may have had some doubts about the American ability to sustain such an immense undertaking, they had no idea that technology would do anything other than promote the republican cause. They cast inventors in the role of dutiful son, the good citizen advancing the public welfare. Their passionate appeals paralleled those of manufacturing enthusiasts. All singled out the inventor for notice, especially manufacturing enthusiasts who tried to show how invention, technological innovation, and manufacturing could secure national prosperity. And so inventors became heroic citizens, at least in their own eyes and in the eyes of later generations of Americans who thrilled to the exploits of Fulton and Edison.

NOTES

1. For the problem of defining "Yankee ingenuity" and the development of a peculiar "American system" of manufactures, see Frank W. Fox, "The Genesis of American Technology: 1790-1860," *American Studies* 17 (1976):29-48; Eugene S. Ferguson, "On the Origin and Development of American Mechanical Know-How,' " *Mid-Continent American Studies Journal* 3 (1962):3-15; and John E. Sawyer, "The Social Basis of the American System of Manufacturing," *Journal of Economic History* 14 (1954):361-379. Also see Dirk Struik's Marxist interpretation of the meaning of political independence and the democratization of science in his *Yankee Science*, pp. 39-40.

2. Reprinted in the *United States Magazine* I (January 1779):23-24, 25.

3. Timothy Matlack, *An Oration* (Philadelphia: Styner and Cist, 1780).

4. Owen Biddle, *An Oration* (Philadelphia: Francis Bailey, 1781), p. 30.

5. Thomas Bond, *Anniversary Oration* (Philadelphia: John Dunlap, 1782), pp. 26-28, 30.

6. Francis Hopkinson, *The Miscellaneous Essays and Occasional Writings of Francis Hopkinson*, 3 vols. (Philadelphia: T. Dobson, 1792), I, 359-360; also see Hopkinson to Franklin, letter of 24 May 1784 in the Franklin Papers, XXXI, fo. 185, APS MSS. Hopkinson, Biddle, Bond, and Matlack were only four from among a great body of orators, politicians, and poets waxing eloquent on this theme. See for example Timothy Dwight, *Greenfield Hill*, Book VII, pp. 161-162, for enthusiastic couplets on American scientific prowess and promise.

7. *An Act to Incorporate and Establish a Society for the Cultivation and Promotion of Arts and Science* (Boston: Benjamin Edes & Sons, 1780); also Bates, *Scientific Societies*, pp. 9-11; and Hindle, *Pursuit of Science*, pp. 263-265.

8. James Bowdoin, *A Philosophical Discourse* (Boston: Benjamin Edes & Sons, 1780), pp. 8-29. Connecticut formed a similar organization in 1786, five years after Ezra Stiles first tried to get it chartered. It failed after a few months but appeared again in 1799 as the Connecticut Academy of Arts and Sciences. See Bates, *Scientific Societies*, pp. 14-15; and Hindle, *Pursuit of Science*, pp. 266-277.

9. *American Academy of Arts and Sciences Memoirs*, I (1785):iii-iv. One of the first inventions communicated to the society was a fire engine pump designed by Saratoga veteran Henry Dearborn, described on pp. 520-524.

10. *Conn. Hist. Soc. Coll.*, XXIII, 204.

11. South Carolina in John F. Grimke, ed., *The Public Laws of the State of South-*

Carolina (Philadelphia: R. Aitken and Son, 1790), p. 344. Bishop, *Amer. Manufactures*, I, 114, 578, for Donaldson; Wall in *Pa. Gazette*, 16 April 1788. The *American Magazine* 1(January 1788):67, noted that two New Yorkers, one of whom may have been John Stevens, had made a dredge sold to the city of Albany. It operated on the same principle as Donaldson's "Hippopotamus," but without the horse for power.

12. Dahn, "Colonial Patents," p. 348; Greville Bathe and Dorothy Bathe, *Oliver Evans: A Chronicle of Early American Engineering* (Philadelphia: Historical Society of Pennsylvania, 1935), appendix C, p. 289.

13. "Early Proceedings of the APS," pp. 137-140; premium notice in the *Columbian Magazine* 1 (December 1786):178-180.

14. Dexter, ed., *Diary of Ezra Stiles*, II, 440-442, 470-471, 474.

15. *American Magazine* 1 (April 1788):349.

16. APS Archives no. 10, Misc. Communications; and the John Fitch Papers in the Peter Force Collection, Series 8D, Item No. 47, Misc. MSS (Washington, D.C.: Library of Congress, 1977), Correspondence and Related Material, 1784-1789 (reel 84) for Turner and Wells.

17. *Columbian Magazine* 2 (January 1787) for a diagram of a truss bridge; and the December 1787 issue, p. 826, for a diagram of a machine to sow wheat. John Beale Bordley communicated news of an English threshing mill and a new way to reap wheat in the October 1788 issue, pp. 577-578, and September 1788, pp. 510-512, respectively. Notices on inventors are in other magazines, such as the *Massachusetts Magazine* 1 (June 1789):336, for potash; and the *American Magazine* 1 (March 1788):271, for a Philadelphia clock inventor.

18. For example the *Boston Magazine* carried pieces on the AAAS but had more articles on flying squirrels and snakes than technologically related items. The only invention noted in three years was an American-made balloon equipped with sails, a drawing of which is in the February 1784 issue, pp. 147-148.

19. Llewellyn Edwards, "The Evolution of Early American Bridges" *Newcomen Society Transactions* 13 (1932-1933):95-116.

20. A description and sketch of the bridge are in the *Massachusetts Magazine* 1 (September 1789):533-534; a note on its opening is in the *Independent Gazetteer*, 1 July 1786, for Cox see Boyd, ed., *Papers of Jefferson*, XVI, 579-580; and Katherine W. Clendenning, "Lemuel Cox," in Malone, ed., *Dict. Amer. Bio.*, IV, 479-480. Cox also invented a machine for cutting wool card wire.

21. Malden Bridge in the *Massachusetts Magazine* 2 (September 1790): 515-516; and the Essex Bridge in Robert Rantoul, "The Building of the Essex Bridge," *Essex Institute Historical Collections* 30 (1893):53-105.

22. Boyd, ed., *Papers of Jefferson*, XIII, 588; Foner, ed., *Writings of Thomas Paine*, II, 1026-1028, 1258, 1266-1267, 1291, 1303, 1321; and the general account in David F. Hawke, *Paine* (New York: Harper and Row, 1974), pp. 163-165, 195, 212.

23. Bathe and Bathe, *Oliver Evans*, pp. 1-8; praise for Evans in Bishop, *Amer. Manufactures*, I, 147, 149, 150. Also see Eugene Ferguson, *Oliver Evans* (Wilmington: Eleutherian Mills-Hagley Foundation, 1977), passim.

24. Bathe and Bathe, *Oliver Evans*, p. xvii; also Watson, *Annals of Philadelphia*, II, 454-455.

25. Bathe and Bathe, *Oliver Evans*, pp. 15-20. New Hampshire's seven-year

patent had a restriction that in practice almost nullified it. Evans asked for a twenty-five-year rather than a fifteen-year patent in Delaware, but to no avail. Martha Ellicott Tyson, *A Brief Account of the Settlement of Ellicott's Mills, With Fragments of History Therewith Connected* (Baltimore: Maryland Historical Society, 1871), pp. 38-39, argued that the Ellicott brothers, not Evans, should receive credit for "inventing" the automated mill. According to her, Joseph Ellicott had been using elevators and hopper-boys as early as 1761, before he emigrated from Bucks County, Pennsylvania. A nineteen-page manuscript from the Ellicott Family Papers (author unidentified), a copy of which is owned by Silvio A. Bedini and was loaned to me, says much the same thing. Yet a typescript copy of John Tyson's "The Rise and Progress of the Town of Ellicotts' Mills. The Founders of Ellicotts' Mills," dated 15 May 1847 and also in the collection owned by Mr. Bedini, states otherwise. George Terry Sharrer, "Flour Milling and the Growth of Baltimore, 1783-1830," (Ph.D. dissertation, University of Maryland, 1975), pp. 65-89, concedes that the Ellicotts did make important contributions, but he also points out that Evans "incorporated new power applications, distributed mechanical stress according to mathematical principles, and arranged several non-continuous operations into a continuous sequence" (p. 72)—and thus Evans deserves his reputation. Sharrer, I believe, makes a better case for Evans than Martha Tyson does for her ancestors, the Ellicotts. Interestingly enough, four Ellicotts were listed as subscribers to the Evans method in a broadside, *Improvements on the Art of Manufacturing Grain into Flour or Meal* (Philadelphia: William Young, 1791), and at least three were authorized by Evans to issue licenses for building his mills. Evans and a largely later generation of Ellicotts suffered a falling out by 1815. Evans brought suit against them and won.

26. Bathe and Bathe, *Oliver Evans*, pp. 11-12, 18, 20, 21.

27. Ibid., p. 16.

28. Six Pennsylvania patents, granted from 1785 to 1787, are listed in the George Bryan Papers, Box 3, fo. 17,, HSP MSS.

29. PCC, Item 41, VI, 490-491 (reel 51); Item 180, 59 (reel 196).

30. Ibid., Item 41, VI, 504-506, 508 (reel 51).

31. Henry Knox Papers, XXVIII, 96 (reel 28); also XLVII, 180 (reel 47).

32. See the PCC, Item 42, I, 335-337 (reel 53); Item 32, 579 (reel 39) for Michael Byrne's device for measuring latitude in 1785. This was probably the same Byrne (called "Daniel Byrne") who contacted the APS in June 1788 to claim that he could read longitude by lunar observation; see "Early Proceedings of the APS," p. 161. One of the most persistent inventors to approach Congress was John Churchman who from 1787 to 1792 sought the aid of Congress, the APS, George Washington, and Thomas Jefferson at one time or another. He too claimed to have discovered a precise way to measure longitude (magnetic variation—which John Macpherson said Churchman stole from him). Churchman attracted some notice in Europe but little here. In 1791 he tried unsuccessfully to get Congress to award him $3,000 for a Baffin's Bay expedition to prove his theory. See Boyd, ed., *Papers of Jefferson*, XV, 139-140, 439-440; XVIII, 61, 492; "Early Proceedings of the APS," pp. 148-150; and *Annals*, 1st Congress, 1st session, pp. 149, 178. Also see the general account in Hindle, *Pursuit of Science*, pp. 349-352. In May 1789 Churchman wrote a plan for a national scientific society to be called the "Washington Society of Sciences," which he suggested meet monthly in New

York City (then the capital). Nothing came of his idea. See his letter and proposal to George Washington in the GWP, 7 May 1789, series 4 (reel 100).

33. Boyd, ed., *Papers of Jefferson*, XI, 293-294, 561-562.

34. PCC, Item 32, 265 (reel 39); also Ford, ed., *JCC*, XXI, 1041.

35. PCC, Item 32, 475-490 (reel 39); Item 19, V, 211 (reel 28).

36. Colles to Congress in ibid., Item 42, II, 168 (reel 53); to Washington in the GWP, letter of 17 January 1783, series 4 (reel 89); Washington's discouraging reply in Fitzpatrick, ed., *Writings of Washington*, XXVI, 64-65.

37. Christopher Colles, *Proposals for the Speedy Settlement of the Waste & Unappropriated Lands* (New York: Samuel London, 1785); the future Erie Canal followed Colles's route to Rome, where it then veered farther south and emptied into Lake Erie instead of Lake Ontario. Also see Ristow, ed., *Survey of U.S. Roads*, pp. 36-52.

38. *Annals*, 1st Congress, 1st session, p. 701. Colles died in New York City in October 1816, having made some influential friends and a certain amount of fame, though no fortune.

39. Ibid., pp. 149, 170-172, 178, 233, 266, 343, 642, 658, 688; also Hazard, ed., *Pa. Col. Rec.*, XVI, 88, 346, for Briggs's application to Pennsylvania.

40. Max Farrand, ed., *The Records of the Federal Convention*, 4 vols. (New Haven: Yale University Press, 1966), II, 325, 505, 616, 620, 655. Also see A. Hunter Dupree, *Science in the Federal Government* (Cambridge: Harvard University Press, 1957), pp. 3-19; Hindle, *Emulation and Invention*, pp. 17-20; Prager, "Historic Background and Foundation of American Patent Law," pp. 317-320. Also see *Annals*, 1st Congress, 1st session, p. 173, for Thomas Tucker's argument that Congress should draw the line of help to inventors at the patent act, and go no further.

41. Hindle, *Emulation and Invention*, p. 25.

42. PCC, Item 19, IV, 39-42 (reel 27); McMechen's application in ibid., Item 78, XVI, 359-363 (reel 99). Perhaps the federal government would not make a grant at that time because it did not hold clear title to the land west of Pennsylvania and Virginia until March 1784. For an overview of the Fitch-Rumsey controversy, see James T. Flexner, *Steamboats Come True*, passim; Hindle, *Emulation and Invention*, pp. 25-57; Frank D. Prager, "The Steamboat Pioneers Before the Founding Fathers," *Patent Office Society Journal* 37 (1955):486-522, and idem, "The Steamboat Interference, 1787-1793," in ibid. 40 (1958):611-643. Also see H. A. Gosnell, "The First American Steamboat: James Rumsey Its Inventor, not John Fitch," *Virginia Magazine of History and Biography* 40 (1932):14, 22, 124-132. The John Fitch Papers, Correspondence and Related Material, 1784-1789 (reel 84) contain virtually all of Fitch's letters and publications on the steamboat, as well as many of those of Rumsey. Other parts of the collection contain the Fitch Steamboat Company financial papers and surviving sketches.

43. Rumsey's petition is in the PCC, Item 42, VI, 498-499 (reel 55); the favorable committee report is in May 1785 in ibid., Item 19, V, 271-272 (reel 28); and Ford, ed., *JCC*, XXVIII, 349-350. The connection between McMechen and Rumsey has not been adequately explained.

44. Virginia patent in Hening, ed., *Statutes at Large*, XI, 502.

45. John Fitch, "Steamboat History," in Prager, ed., *Autobiography*, p. 143.

46. The Fitch application is in the PCC, Item 78, IX, 544 (reel 95). The letters

of support were from John Ewing, provost of the University of Pennsylvania, and Samuel Smith and William Houston of the College of New Jersey. Congress's dismissal of the petition is in Ford, ed., *JCC,* XXIX, 669n, 672n, 673n.

47. Fitch, "Steamboat History," in Prager, ed., *Autobiography,* pp. 150, 153.

48. "Early Proceedings of the APS," pp. 133, 135.

49. PCC, Item 42, III, 131 (reel 54); Ford, ed., *JCC,* XXX, 158n.

50. Fitch, "Steamboat History" in Prager, ed., *Autobiography,* pp. 156-157, 166-167. Donaldson had the support of John Stevens—see Turnbull, *John Stevens,* p. 104.

51. Hening, ed., *Statutes at Large,* XII, 616-617, for the Fitch patent.

52. Fitch, "Steamboat History," in Prager, ed., *Autobiography,* pp. 167-168.

53. Ibid., p. 185.

54. PCC, Item 19, II, 277-278 (reel 26) for the committee decision of 4 March 1788. Fitch's application is in ibid., Item 42, III, 169-170 (reel 54).

55. Rumsey to Washington, 24 March 1788 in the GWP, series 4 (reel 97); also James A. Padgett, ed., "Letters of James Rumsey," *Maryland Historical Magazine* 32 (1937):141-142. Also see the letter from Joseph Clay to George Turner of 29 March 1789, APS Misc. MSS. Coll., on Georgia's delay in granting Rumsey a patent.

56. Rumsey to Washington, 15 May 1788, in ibid., series 4 (reel 97); also a circular on the Rumseian society in ibid., series 4, under 1 May 1785 (reel 97). Also see Padgett, ed., "Letters of James Rumsey," pp. 147-150; and letter of Miers Fisher to Robert Barclay of 5 April 1790 in the APS archives, on how the Rumseian society aided its founder.

57. Rumsey had by then also devised ways of applying steam to raising water and to grist mills and sawmills, all of which Pennsylvania and New York patented. See his *Explanations and Annexed Plates of the Following Improvements in Mechanics* (Philadelphia: Joseph James, 1788) and his appended *Explanation of a Steam Engine.* See the *Pa. Packet,* 4 January 1790, for license advertisements for this machinery.

58. "Early Proceedings of the APS," p. 172.

59. James Rumsey, *A Short Treatise on the Application of Steam* (Philadelphia: Joseph James, 1788); John Fitch, *The Original Steamboat Supported or a Reply to James Rumsey's Pamphlet* (Philadelphia: Zachariah Poulson, Jr., 1788).

60. Joseph Barnes, *Remarks on Mr. John Fitch's Reply to Mr. James Rumsey* (Philadelphia: Joseph James, 1788). Also see the comments of John Vaughan in letters to his brother Benjamin of 15 and 24 May and 4 June 1788 in the APS Misc. MSS. Coll. John Vaughan was a member of the Rumseian society, while his father, Samuel, was a member of Fitch's company. John Dickinson supported Fitch. See his letter to the Pennsylvania Legislative Council of 19 January 1787, LC Misc. MSS.

61. Boyd, ed., *Papers of Jefferson,* XV, 145. Perhaps Rumsey took solace that his improvements in mill machinery were being used at a gunpowder factory near Philadelphia. See William Barton, "Remarks on the state of American manufactures and commerce," *American Museum* 7 (May 1790):289.

62. Letter to Franklin of 12 March 1788 in the Franklin Papers, XXXVI, fo. 33, APS MSS.

63. See the list of patents from 1790 to 1804 prepared by Secretary of State

James Madison and reprinted in Thomas C. Cochran et al., eds., *The New American State Papers: Science and Technology*, 14 vols. (Wilmington: Scholarly Reprints, 1973), IV, 25-33.

64. *Annals*, 1st Congress, 1st session, p. 335.

65. Boyd, ed., *Papers of Jefferson*, XV, 170-172.

66. Ibid., XIV, 699.

67. Compare Jefferson's letters to Madison of 31 July 1788 and 28 August 1789 in ibid., XIII, 443; and XV, 368, respectively.

68. Hugo Meier, "Thomas Jefferson and a Democratic Technology," in Carroll W. Pursell, Jr., ed., *Technology in America* (Cambridge: MIT Press, 1981), pp. 17-33.

69. Fitzpatrick, ed., *Writings of Washington*, XXX, 493.

70. See George Clymer to Tench Coxe in the Tench Coxe Papers, series II (reel 55), 1790; quote from a letter to James Madison of 21 March 1790 as cited in Davis, *Earlier History of Amer. Corp.*, I, 355.

71. *American Magazine* 1 (January 1788):67. For Webster's role in bringing that legislation, see Frank D. Prager, "Proposals for the Patent Act of 1790," *Patent Office Society Journal* 36 (1954):157-158.

72. Hindle, *Emulation and Invention*, p. 57.

73. *Annals*, 1st Congress, appendix, pp. 2270-2273. Also see P. J. Federico, "Operation of the Patent Act of 1790," *Patent Office Society Journal* 18 (1936):237-251; and State Department clerk Henry Remsen's note on procedures in Boyd, ed., *Papers of Jefferson*, XVII, 384-385. Also see Prager, "Proposals for the Patent Act of 1790," pp. 157-167, for the differing views of Fitch and Rumsey.

74. Boyd, ed., *Papers of Jefferson*, XVI, 579. On January 30, 1789, the Delaware Assembly had voted to let Congress settle the steamboat dispute. See the Fitch Papers, Correspondence and Related Material, 1784-1789, (reel 84).

75. Cochran, ed., *New Amer. State Papers: Science and Technology*, IV, 16-18; also Fitch, "Steamboat History," in Prager, ed., *Autobiography*, pp. 197-198. Fitch never understood the legislation—see his letter to John Nicholson of 3 January 1792 in the Gratz Collection, Case 7, Box 35, HSP MSS., where he expressed hope for securing a twenty-five-year patent and land from Congress.

76. Letter to Levi Hollingsworth of 30 June 1790 in the APS Misc. MSS. Coll.

77. Cochran, ed., *New Amer. State Papers: Science and Technology*, IV, 11-12, 14-15.

78. Joseph Barnes, *Treatise on the Justice, Policy and Utility of Establishing an Effectual System for Promoting Progress of the Useful Arts* (Philadelphia: Francis Bailey, 1792), p. 9.

79. *Annals*, 2nd Congress, appendix, pp. 1431-1435.

80. Oliver Evans, *The Young Mill-Wright and Miller's Guide*, 3rd edition (Philadelphia: M. Carey and Son, 1818), appendix, p. 363.

81. Fitch Papers, Correspondence and Related Material, 1784-1789, letter of 10 April 1788 (reel 84). Arthur Donaldson noted in a petition to the Pennsylvania Assembly of 18 March 1786 that his Hippopotamus, though patented, earned him no money. See ibid. Fitch wrote this lament in 1792, also in ibid:

> For full the scope of seven years
> Steam Boats excited hopes and fears

In me, but now I see it plain
All further progress is in vain
And am resolved to quit a Scheming
And be no longer of pattents dreaming
As for my partners *Dam them all*
They took me up to let me fall
For when my Scheme was near perfection
It proved abortive by their defection
They let it stop for want of Rhines
Then swore the cause of failer mines.

82. Note of 13 February 1790 in ibid. (reel 85).

83. Cochran, ed., *New Amer. State Papers: Science and Technology* IV, 15.

84. Joel Barlow, *The Vision of Columbus* (Hartford: Hudson and Goodwin, 1787), p. 203. Perhaps some were too optimistic, in this and in other ways. See Joseph Ellis, *After the Revolution* (New York: W. W. Norton & Company, 1979).

85. Ezra Stiles, *The United States Elevated to Glory and Honor* (New Haven: Thomas & Samuel Green, 1783), passim; Chastellux, *Travels in North America*, II, 545-546; and the astute predictions in Thomas Pownall, *A Memorial Most Humbly Addressed to the Sovereigns of Europe, on the Present State of Affairs between the Old and New World* (London: J. Almon, 1780), pp. 35–56, 69–70.

9 INVENTING A NATION

Revolutionary America underwent a mechanical metamorphosis, a change in technological attitudes that anticipated a later change in technological prowess. The technological aspirations of Americans in 1760 had grown considerably by 1790. By then the new nation had begun to pursue technological independence almost as vigorously as it pursued political independence. For many Americans, the two were inseparable. Technology was an essential part of their ideal of the virtuous republic.[1] Americans searched for evidence that they were building that republic and fulfilling their national mission, and they settled on technology as proof that they were. Technology sustained their utopian dreams.

Their technological revolution, like their political revolution, took them in new directions and led them to unforeseen consequences.[2] If Americans were still technologically innocent, most wanted to become more technologically worldly. They went busily about creating a technologically oriented nation, not knowing exactly what it would bring. Most viewed the technological future optimistically. Some of their projections were overly optimistic, if not for the quickening rate of technological change, then for the amount of good that that change necessarily carried with it.

It was no accident that the Revolutionary generation built a technologically oriented society. In 1760 the colonies were part of a mercantilistic empire that forbade certain forms of technological endeavor while it encouraged others. Many colonists protested against that empire and condemned it as a symbol of British repressiveness. Colonial legislatures could legally support only specified technologies, they complained, technologies that only incidentally benefited them. The Pennsylvania Assembly therefore acted illegally when it promoted the manufacture of steel. Other colonial assemblies often did the same when they pushed for home manufactures, so their attempts to manufacture brought into sharp relief what were once only vague frustrations with the empire.

Some years before the war erupted, John Dickinson had admonished in his *Letters from a Farmer in Pennsylvania,* "let us all be united with one spirit, in one cause. Let us invent—let us work—let us save."[3] Dickinson called for an American effort to counter what he saw as Parliamentary excess and, like other critics of mercantilism, he emphasized the need for a more balanced, more vibrant American economy. Invention was not promoted as vigorously in 1767, when Dickinson wrote the lines above, as it would be twenty years later by Tench Coxe, just as, in 1767, as least, John Dickinson was not yet an advocate of political independence. Dickinson eventually if reluctantly swam with the revolutionary tide. Invention too was carried along, swept up in the public consciousness-raising known as building a national identity.

Colonial political institutions were either reorganized or replaced after the War of Independence. Post-war politics and government were better suited to American technological ambitions. The Pennsylvania state legislature could act in concert, not at cross-purposes, with the national government if it chose to promote technological change. The same could be said of the national government and the American economy after the political reorganization under the Constitution in 1789, when the navigation acts of the colonial period were replaced with the import duties of the early national era. Navigation acts were designed to circumscribe the colonial economy, import duties were intended to promote industrialization. Most technological enthusiasts—especially those interested in manufactures—had pushed for ratification of the Constitution, just as they soon after pushed for federal patent laws as well as tariffs. All were seen as part of a larger nationalistic program, a program designed to secure technological independence.

Americans had gone to war to determine their own technological as well as their own political destiny. That war was waged by a society in transition, midway between economic dependence and economic self-sufficiency, poised on the threshold of industrialization. Industrialization after the war brought a greater push for technological freedom and with it a greater willingness to pursue technological innovation.

The War of Independence itself helped to shape the course of what followed. Eight years of economic privation threw a scare into Americans that they did not soon forget. The war laid bare their technological shortcomings. In desperation they built a munitions industry; in desperation they turned to the Pennsylvania rifle; in desperation a few even looked to David Bushnell. Bushnell could work no miracles, and neither could riflemen; munitions were always in short supply. Americans survived the war anyway. Even if they did not do well with devices like the "Turtle" or tools like the rifle, they believed that they had won the war because of their superior perseverance and ingenuity. Some grudgingly gave the French credit, but most preferred to congratulate themselves. They saw themselves as a people who had used their wits to overcome all odds. And indeed, wartime desperation had put a premium on wit—on innovativeness. Wartime improvisation opened the door for ingenuity wider than it had been opened during the colonial period, and there was ingenuity enough to be had. If Americans were not ready to capitalize on the ingenuity of David Bushnell, they were perfectly able to draw on the ingenuity of Robert Smith.

Artisans like Smith performed journeyman wartime service. Politically, craftsmen after the war began to taste the power of representative democracy; technologically, they enjoyed more opportunity as well. Postwar industrialization was based on the machines they designed and introduced as technological innovations. Americans had for years thought of themselves as an ingenious people, and they eventually came to equate that ingenuity with mechanical inventiveness. The inventor was viewed as simply the most ingenious citizen in a republic of the ingenious.

Americans paired technology with republican notions, yet they were not always of one mind when it came to the promotion of individual technologies. The technological priorities of a manufacturing enthusiast could be quite different from those of a handicraft-based artisan. To characterize Revolutionary Americans as "inventive" or "ingenious" is deceiving unless those terms are qualified. To be sure, the new nation abounded with ingenious individuals, but Americans did not always rush to discard the old for the new. For every Oliver Evans, there were ten millers reluctant to automate; for every George Washington, there were fifty farmers uninterested in scientific farming.

Millers who chose not to adopt the Evans method were not necessarily technologically unimaginative or opponents of technological change. Their commitment to milling may have run counter to the interests represented by automation. Millers who opted to grind their flour the traditional way may have had no use for Evans's machinery. If that machinery made production more efficient and made possible greater profits, it also disrupted a miller's routine and displaced workers, workers who were often kinsmen. The acceptance or rejection of a new technology, then, was not always a question of the technologically informed versus the technologically uninformed, a conflict between the imaginative and unimaginative.

Yet if Evans's career shows how large-scale technological change was seldom swift in Revolutionary America, it also illustrates what was ultimately the irresistible attraction of automated milling. Many of those millers who waited a decade to adopt the Evans method already produced grain for market. Since they were involved in commerce and concerned with making a profit, perhaps it was inevitable that they would change. Their desire for even greater commerce and higher profits may have overcome their suspicions of the "newfangled" and their attachment to a partially mechanized handicraft trade. Linking their conversion to automation with the national quest for technological independence may also have made them feel that they were contributing to community health. Perhaps some were fatalists and simply reconciled themselves to a change they could not prevent: if one miller changed, they may have concluded, they would all have to if they wanted to remain competitive. Either way, whether they embraced automated milling enthusiastically or reluctantly, their experience was a microcosm of the larger American experience with mechanization.

Mechanization, of course, had not begun suddenly with the Revolutionary Era nor did it follow simply because of it. The Colonial Era had not been part of some pre-technological age; tools had been on the American scene from the beginning. Still, many of those tools had not changed markedly over the years. The new machines designed or imported for the textiles industry in the 1780s and 1790s, like the devices used in sawmills before the war, cannot be used to gauge the overall rate of change. More important than the replacement of old machines with new machines in the Revolutionary Era was the growing public fascination with the process of change itself and the popular identification with technology as a vital—and positive—social force.

Now this does not mean that a national commitment to technological progress always accelerated the actual rate of change. Making such a cause and effect relationship would be extremely difficult, requiring a venture into counterfactual history—that is, speculating on what would have happened to American technology if no one had ever made any public statement about it. Yet it does seem logical that the public cry for technological change made that change just so much more attractive. It may have caught up inventors who otherwise might not have invented, filling them with patriotic rationales for their attempts to make money off their genius. Likewise with merchants and bankers who backed new technologies. Perhaps public enthusiasm for technology simply hastened a commitment to investment and development that would have come anyway, but that commitment would have come without the attendant justifications and pairing of technology with republicanism. It was that pairing that helped define national purpose, giving specific form to what otherwise might have remained a vague faith in the idea of progress.

The mechanical metamorphosis of Revolutionary America was part of a larger transformation of American society, a transformation that reinforced the quest for material progress. That transformation had begun much earlier and extended far beyond the Revolutionary Era, a movement that Richard D. Brown described as the shift from a "traditional" to a "modern" society. According to Brown, the American Revolution shattered "the stability and balance that evolved between modern and traditional forces." The Revolution advanced the cause of modernization, and technological change was an agent of modernization. Nevertheless, as Brown concedes, the movement from a traditional to a modern society did not proceed smoothly; movement came in spurts, and not always in the same direction. There was uneven progress, and there were tendencies that rolled on in "quiet contradiction."[4] Besides, Brown would probably agree that distinctions between "traditional" and "modern" are not easy to make and even more difficult to apply to a particular time or place. To speak of something as nebulous as shifting values or national will is to risk overgeneralization. Nonetheless, to avoid generalization altogether is to avoid any attempt to explain the essence of the American experience.

True enough, Americans often disagreed over which technologies should be encouraged. Hardly any, however, opposed technological change *in toto*. There was no need, they decided, because their political creed would keep technology from becoming abusive. Their virtuous republic, they believed, would prevent any festering sores from appearing on the national landscape. There would be no Manchesters in America; there would be no industrial blight. Most Revolutionary Americans consequently embraced manufactures without understanding that the type of industrialization that followed would transform the very society they had expected manufactures to protect. Not seeing that far into the future, they had few misgivings.[5]

Admittedly, there were those among the Revolutionary generation who objected to certain types of manufacturing enterprise. George Logan and Noah Webster found themselves on opposite sides of the political parties that emerged by the 1790s, but they were of one mind when it came to the dangers of large-scale manufacturing. Yet if Logan was opposed to manufactures that could cause the creation of an urban laboring class, he was a supporter of smaller scale operations. If Webster was opposed to heavy manufactures for similar reasons, he was also a defender of the patent system and rabid promoter of things "American." Logan and Webster were ambivalent at best, as were other opponents of large-scale technological change. They had to juggle their fears of what the British had done and what they themselves might do with their hopes for the new republic. By the time that these opponents to large-scale manufactures closed ranks and made clear their position on what should be encouraged and what should not, events had left them behind. A national technological commitment had been made and, once made, it was difficult to qualify.

Southern agrarians would someday rail against northern industrialists, but those same agrarians were devoted to a cotton culture based in part on Eli Whitney's gin. They wanted only those technologies that complemented their own peculiar brand of capitalism, but they staged no technological revolts. Even Thomas Jefferson, suspicious of the new technological order looming ahead, created for himself a philosophical compromise Leo Marx has called the "middle landscape."[6] If Jefferson thought that localized household manufactures could be fostered in America and heavy industry be kept in Europe, he was as wrong as those in the 1760s who believed that colonial manufactures could be reconciled with the empire. After all, Jefferson, the inveterate tinkerer, made no attempt to disguise his excitement over steam engines. Those same steam engines supplied power to the British factories that he feared would pollute his bucolic paradise if brought across the Atlantic. The material progress Americans pursued would not stop at the conflicting limits set up by Jefferson or, some years before, by the Board of Trade. Just as moderates like Dulany had played into the hands of radicals like Rush before the War of Independence, Jefferson played into the hands of Hamilton and Coxe in the post-war years.

Industrialization marched on. The veritable flood of British artisans entering the country after the War of Independence, in a well-known story recounted briefly earlier, boosted American industrial expertise. A little-emphasized footnote to that story is that the arrival of British artisans and the transfer of British ideas and machines after the War of Independence assisted native-born inventors.

The pre-war home manufactures movement had done little to awaken Americans to the latent power of invention. The post-war movement, by drawing more directly on the Industrial Revolution, changed all of that. For indeed, the more directly the United States entered the industrial world, the more the fortunes of inventors rose. Oliver Evans tried and failed for ten years to persuade his neighbors in the Brandywine district to adopt his machinery. In the mid-1790s, after the obsession with textile machinery imported from Britain stimulated an obsession with machines in general, more and more mills changed over to the Evans method. That change, it would seem, was too striking to have been purely coincidental. Once largely ignored, Evans had to scramble to stay ahead of other inventors who were hoping to improve on his ideas.

In an appendix to his *The Young Mill-Wright and Miller's Guide*, Evans penned his thoughts on "The True Paths to Inventions." Foreshadowing the closer link of science and technology in the nineteenth century, he broke invention down into four consecutive steps: investigating fundamental principles, choosing their best application, deciding whether current machines were adequate, and putting into practical use the discoveries made.[7] Evans saw invention as a systematic process, subject to procedural rules. He anticipated the day when invention would become less a product of random trial and error and more a designing of new devices to fit previously researched needs. Thus he explained to readers his desire to join theory and practice to make the art of milling a science.

Evans's magnified vision came simultaneously with the maturation of a national technological aptitude. New machines and new processes were pulled more directly into the life of the new nation because the economy, government, politics, and culture had combined to create an atmosphere more conducive to technological change. Industrial experiments highlighted by the success at Pawtucket were the natural result. Inventors made new devices which, if introduced as technological innovations, promised rewards to inventor and investor alike. Those rewards came when society had reached a state where inventions and the goods they produced could be marketed. A system of mutual reinforcement evolved. Society stimulated the inventor, the inventor responded with new devices, the new devices were incorporated into the economy, and society took another technological step forward. The process could then be repeated. That process would move the textiles industry from its modest beginnings at Pawtucket to the great enterprises at Lowell.

The industrial growth evident by the 1790s had a marked effect on "Yankee ingenuity." Inventors like Oliver Evans found more outlets for their genius and, with the promise of greater profits, more incentive to invent. H. J. Habakkuk contended that "to a considerable extent American ingenuity was a result rather than a cause of mechanization." He therefore dated the appearance of an identifiable "Yankee ingenuity" from the 1820s rather than the 1790s.[8] Habakkuk's conclusion rings true if ingenuity is defined in narrow terms. Habakkuk defined ingenuity as the production of laborsaving industrial devices. He reviewed the history of technological innovation in the Revolutionary Era and concluded that American industrial "know-how" was not evident until well after the machine tool and textile experiments of Eli Whitney and Samuel Slater. In the sense that American industrial "know-how" came after the adoption of British mechanical skills and new industrial devices, Habakkuk would appear to be right. But perhaps Habakkuk's definition needs to be broadened. The ingenuity associated with manufacturing had been preceded by a different type devoted to the design of useful if not industrially significant tools. Habakkuk's Yankee, the industrial innovator, had been preceded by another Yankee, the tinkerer. Moreover, the Habakkuk model does not adequately explain the genius of Oliver Evans.

Like Habakkuk, popularizers of the "ingenious Yankee" have not distinguished between the two types of inventive activity. Roger Burlingame inconsistently discussed the collective ingenuity of early Americans on the one hand and chronicled the shattered careers of individual inventors on the other. His theme would have been less strained had he allowed for the evolutionary process that carried inventors from the workshops of Colonial America to the Menlo Park laboratory of Thomas Edison. Burlingame correctly identified "Yankee ingenuity" as part of even the earliest American experience, but those early ingenious Yankees worked within the constraints of a somewhat less dynamic society. Change in the social function of invention came with the Industrial Revolution and with the American quest for political and technological independence. As it stands, unchanged to allow for the formative influence of British ideas and machines, Burlingame's brand of Turnerianism is misplaced.

The tinkerer celebrated by Burlingame did not disappear with industrialization. The tinkerer survived, not only as a flesh and blood inventive jack-of-all-trades, but as a folk hero. Thomas Edison is the most beloved of the nation's inventor heroes because he seemed to be both a tinkerer and an industrial innovator. His fame rests on modern American technological ambivalence, on nostalgia for the simple pre-industrial past as well as enthusiasm for the complex "post-industrial" future.

As Eugene Ferguson noted some time ago, reference to "Yankee ingenuity" does not explain the process by which Americans emerged as a technologically oriented people by the mid- nineteenth century.[9] Still, a proper understanding of it is essential to understanding American technological development. Promotion of invention and ingenuity was part of a maturing technological outlook observable by the end of the Revolutionary Era, an essential element of the new technological order Americans wanted to create.

By 1790 Americans were more sure of themselves, politically and technologically. Samuel Slater, the Englishman, had teamed with Moses Brown, the Yankee, to help inaugurate a textiles revolution in New England. Eli Whitney's cotton gin would soon reinvigorate the slave economy of the South and indirectly further that revolution. By the early decades of the next century, Oliver Evans lived to see his inventions used on western rivers as well as in eastern flour mills, and he had the satisfaction of seeing inventors taken more seriously.

The evolution of the patent system from the 1780s through 1790 reflected a growing public appreciation for the social importance of people like Evans. By no longer allowing blanket patents of the type awarded in some states to John Fitch in the 1780s, the federal code better defined the limits of intellectual property. Inventions were seen as a source of national wealth, the property of society as well as the individual. Relatively few advocates of federal patent legislation had been disinterested parties, motivated solely by a selfless concern for science, technology, and social welfare. Altruistic or not, their actions flowed from a realization that invention could be a source of economic power. After the Revolution, inventors of a serious bent, with an idea that could be marketed and exploited with a little entrepreneurial imagination, found listeners among Americans who took pride in their rising technological independence.

This is not to say that Americans of the Revolutionary generation fully understood the impact of technology on society or foresaw the potential clash between humanistic and materialistic values that technology brought with it. Grasp of the market end of technological growth—production, profits, and consolidation—advanced at a much faster rate than sensitivity to less desirable side effects. Indeed, some Americans are only now beginning to realize that technology, depending on how it is used, can imprison as well as liberate. Most Americans before the last few decades were too easily hypnotized by technological growth because it complemented ingrained expansionist and increasingly materialistic urges.

Americans are still coming to grips with their two revolutions. The Declaration of Independence and its symbolic defense of basic freedoms has served as a beacon of the political revolution, but Americans have yet to live up to the creed spelled out in that document. If Americans have fallen short there, they have succeeded all too well in living up to their technological commitment. A growing number regret that the two revolutions were ever paired. They would like to turn back the clock but have found that the technological commitment has become stamped on the national character. To question the American technological commitment is to question the very nature of the American experiment. Most are loathe to do that even if they are disillusioned with modern technology.

For good or ill, the technological obsessions of modern America have their public origins in the Revolutionary Era. By 1790 the United States, though in most ways still a nation of farmers, was drawn inexorably into the industrial vortex and reveled in its discovery of the transforming power of technology.[10] Invention became pursued as an economic commodity, just as manufacturing became worshipped as a social boon. As many members of the Revolutionary generation saw it, they could reap the rewards of a technological order and ward off its defects with their republican creed; they could eat their cake and have it too. In language reminiscent of John Fitch, Jonathan Maxcy, president of the College of Rhode Island, told the Providence Association of Mechanics in 1795 that "inventors and improvers of Arts" had "meliorated the condition of society"; inventors, not politicians and generals, had built civilization.[11] Citing Oliver Evans and the spread of his automated mill machinery as proof, one technological enthusiast had editorialized a few years before Maxcy:

> It must afford great pleasure to the friends of American manufactures, to see the rapid improvements which have, within those few years, been made in machinery, within the United States; and too much praise cannot be given to those individuals, who, notwithstanding the numerous obstacles which present themselves against *new inventions* in a young country, by steady perseverance, complete valuable machines.[12]

That editorial appeared in the *American Museum*, a long-time promoter of industrialization and in 1792 a newcomer to the promotion of invention. Like the editors of the *Pennsylvania Gazette* several years before, the editors of this magazine made an inseparable connection between invention, technological change, and national progress.

A Frenchman touring the United States in 1797 took stock of American attitudes and complained that his hosts believed "that the spirit, invention and genius of Europe are in a state of decreptitude . . . whilst the genius of America, full of vigour, is arriving at perfection."[13] He thought this was incongruous, since, as far as he was concerned, the flourishing textile mills at Pawtucket owed their existence to an Englishman. Indeed, one of his countrymen had scoffed sometime earlier that Americans merely adopted machines and processes "that we have discovered and applied before them."[14]

What neither observer could know is that by the time of their visits Samuel Slater had been adopted as a native son. Environmentalistic explanations of his genius overshadowed talk of his national origins. Ignoring the fact that Slater had been preceded by dozens of British artisans who tried but failed to introduce Arkwright machinery into the country, Americans believed that the mere act of emigration had brought a transmutation of talents. They began the myth of an all-encompassing "Yankee ingenuity," too soon forgetting the role of British ideas and British machines in building their new technological republic. Not only did Americans open their eyes to invention, they came to regard inventiveness as their peculiar gift from God. Most proudly identified with Jacob Bigelow's 1816 boast that he lived in a "nation of inventors."[15] Inventors found themselves admired as the geniuses behind technological change, and technological change was revered by many as the great guarantor of a healthy republic.

With each passing decade the American identification with invention and technological innovation became more pronounced. In the gush of national exuberance that marked the centennial in 1876, technology was paid special attention. The year had begun inauspiciously enough. Americans were weary of Reconstruction and recovering from a recession. As the year wore on, they put aside their troubles in a patriotic paroxysm. Millions of visitors flocked to the Centennial Exposition in Philadelphia, there to renew their national pride. To walk through Machinery Hall, which contained the most popular exhibits of the exposition, was to pass in review of the machine as American icon. Machines were monuments to American ingenuity, testaments to the success of the American experiment. "Nothing can be truer than that inventiveness is a *habit* with Americans," wrote one visitor, who no doubt expressed widely shared feelings.[16] As another visitor put it more emphatically, "Our true standing army is one of inventors, not soldiers; and to the former alone, under

God, do we owe our national prosperity."[17] Such were the sentiments of those in 1876 who believed that the true legacy of 1776 was twofold—a new political nation and the beginnings of a technological republic.

NOTES

1. For faith in technology, see Kasson, *Civilizing the Machine;* and Meier, "The Technological Concept in American Social History." For American belief in progress in general and the hope that independence would usher in a new age, see Henry F. May, *The Enlightenment in America* (New York: Oxford University Press, 1976); Nathan O. Hatch, *The Sacred Cause of Liberty* (New Haven: Yale University Press, 1977); and Sacvan Bercovitch, *The American Jeremiad* (Madison: University of Wisconsin Press, 1978).

2. Gordon Wood, *The Creation of the American Republic, 1776-1787* (Chapel Hill: University of North Carolina Press, 1969), p. 389, noted that "men were always only half aware of where their thought was going" when they set out to design their political republic. So too with their technological republic. See James Hiner, "On Distinguishing 'A Machine' from Its System," *American Quarterly* 14 (1962):612-617.

3. John Dickinson, *Letters from a Farmer in Pennsylvania* (Philadelphia: David Hall and William Sellers, 1768), pp. 17-18.

4. Richard D. Brown, *Modernization* (New York: Hill and Wang, 1976), pp. 8, 24, 72.

5. See Thomas Bender, *Toward an Urban Vision* (Lexington: University of Kentucky Press, 1975). McCoy, *The Elusive Republic*, presents a somewhat different view. Kulik, "Beginnings of the Industrial Revolution in America," noted that there was resistance to change in Pawtucket, as indeed there was. Smith, *Harpers Ferry Armory and the New Technology*, p. 23, has concluded that, for some Americans, at least, acceptance of industrialization was "hesitant and equivocal." Nevertheless, that hesitancy and equivocation only became notable after the 1790s, after the development of trends difficult to reverse. American resistance to large-scale technological change, whether in Pawtucket or Harpers Ferry, was never so strong that change was kept from occurring. It may have been impeded, but it was not stopped.

6. Leo Marx, *The Machine in the Garden* (New York: Oxford University Press, 1964), pp. 116-144; see also Jacob Cooke, "The Collaboration of Tench Coxe and Thomas Jefferson," *PMHB* 100 (1976):468-490; and Bruchey, *Roots of American Economic Growth*, pp. 113-122.

7. Oliver Evans, *The Young Mill-Wright and Miller's Guide*, 1st edition (Philadelphia: Printed and sold by the Author, 1795), appendix, p. 1. I have not stressed here what others have discussed in detail elsewhere: the practical, utilitarian bent of American science and technology, which was undoubtedly evident, though not—as some would contend—to the exclusion of more theoretical concerns. For the classic commentary see Alexis de Tocqueville, "Why the Americans are more Addicted to Practical than to Theoretical Science," in *Democracy in America*, 2 vols. (New York: Alfred A. Knopf, 1980), II, 41-47, and in other chapters. Also see David Hume, "Of the Rise and Progress of the Arts

and Sciences," in *Essays and Treatises on Several Subjects*, 2 vols. (Edinburgh: James Clarke, 1809), I, 111-141.

8. H. J. Habakkuk, *American and British Technology in the Nineteenth Century* (Cambridge: Cambridge University Press, 1962), p. 122.

9. Eugene S. Ferguson, "On the Origin and Development of American Mechanical Know-How,' " pp. 3-15; and Fox, "The Genesis of American Technology," pp. 29-48, though Fox emphasized individualism and egalitarianism more than I would. Much-needed words of caution about the individualistic, entrepreneurial Yankee are offered in James A. Henretta, "Families and Farms: *Mentalité* in Pre-Industrial America," *WMQ*, 3rd series, 35 (1978):3-32; Smith, *Harpers Ferry Armory and the New Technology*; and Kulik, "Beginnings of the Industrial Revolution in America," pp. 7-10.

10. See Perry Miller, "The Responsibility of Mind in a Civilization of Machines," *The American Scholar* 31 (1961-1962):51-69, for a brief and incisive discussion of the American love affair with machines; and Hugo A. Meier, "Technology and Democracy, 1800-1860," *Mississippi Valley Historical Review* 43 (1957):618-640. Lewis Mumford contends that the Revolution ironically freed Americans to chain themselves to authoritarian technics, to come under the spell of the megamachine. See his *The Myth of the Machine: The Pentagon of Power* (New York: Harcourt, Brace and Jovanovich, 1970), pp. 24, 41-44.

11. Jonathan Maxcy, *An Oration before the Providence Association of Mechanics* (Providence: Bennett Wheeler, 1795), p. 3.

12. *American Museum* 11 (1792):225-226.

13. As cited in Norman B. Wilkinson, "Brandywine Borrowings from European Technology," *Technology and Culture* 4 (1963):1.

14. Ferdinand M. Bayard, *Travels of a Frenchman in Maryland and Virginia with a Description of Philadelphia and Baltimore in 1791* (Ann Arbor: Edward Brothers, Inc., 1950), pp. 4-5.

15. Jacob Bigelow, "Inaugural Address, Delivered in the Chapel at Cambridge, December 11, 1816," *North American Review* 4 (1817):276.

16. "Machinery," *The Nation* 23 (9 November 1876):283.

17. "American Progress—From 1840 to the Present," *Scientific American* 34 (3 June 1876):353. The piece continued: "It was the inventor who, when the first war for national life left us prostrate though victorious, gave us the means to throw off our dependence on other nations, and stand forth not merely politically but industrially and commercially, a free and independent people."

BIBLIOGRAPHICAL NOTE

I consulted a number of key manuscript and printed document collections while gathering materials for this study. Both the Papers of the Continental Congress, 1774-1789 (Washington, D.C.: Government Printing Office, 1959, 204 reels microfilm) and the George Washington Papers (Washington, D.C.: Government Printing Office, 1965, 124 reels microfilm) provided essential information on various people, from obscure inventors to better-known public personalities, and details on the status of the munitions industry during the war. Large segments of both collections have been published, under Worthington C. Ford, ed., *The Journals of the Continental Congress, 1774-1789*, 34 vols. (Washington, D.C.: Government Printing Office, 1904-1937); and John C. Fitzpatrick, ed., *The Writings of George Washington*, 39 vols. (Washington, D.C.: Government Printing Office, 1932-1945), respectively. Edmund Cody Burnett, ed., *Letters of the Members of the Continental Congress*, 8 vols. (Washington, D.C.: Carnegie Institute, 1921-1936), likewise has much useful material, as does Paul H. Smith, ed., *Letters of the Delegates of the Continental Congress, 1774-1789*, 9 vols. (Washington, D.C.: Library of Congress, 1976-), through May 1778 to date.

Three other microfilm publications of manuscript collections having pertinent information on inventors, technology, and the war are the Papers of Tench Coxe in the Coxe Family Papers (Philadelphia: Historical Society of Pennsylvania, 1977); the Henry Knox Papers (Boston: Massachusetts Historical Society, 1960, 55 reels microfilm); and the Adams Papers (Boston: Massachusetts Historical Society, 1958, 608 reels microfilm). Most of John Adams's writings have been published in Charles Francis Adams, ed., *The Works of John Adams*, 10 vols. (Boston: Little, Brown and Co., 1850-1856). Also see the "Warren-Adams Letters," *Massachusetts Historical Society Collections* 72 and 73 (1917); and the new collection in Robert J. Taylor et al., eds., *Papers of John Adams*, 6 vols. (Cambridge: Harvard University Press, 1977-), which so far carries through to 1778.

One unavoidable obstacle in studying technology in the Revolutionary Era is that the papers of early inventors either have not survived or are dispersed among numerous collections. Nevertheless, the American Philosophical Society and the Historical Society of Pennsylvania are excellent starting points. The Benjamin Franklin Papers at the American Philosophical Society are a virtual treasure trove of materials relating to early American science and technology.

Many of Franklin's letters are printed in Albert H. Smyth, ed., *The Writings of Benjamin Franklin*, 10 vols. (New York: Macmillan Co., 1905-1907); a more complete compilation is underway by Leonard Labaree et al., eds., *The Papers of Benjamin Franklin*, 23 vols. (New Haven: Yale University Press, 1959-), which at present only covers through April 1777. APS manuscript holdings also include miscellaneous communications to the society not printed in the "Early Proceedings of the American Philosophical Society for the Promotion of Useful Knowledge, 1744-1838," *American Philosophical Society Proceedings* 22 (July 1885), part 3, no. 119, which itself is an indispensable collection.

Letters to or from or about James Rumsey, John Fitch, and a few other inventors are located in the APS MSS. collections. The reprinted *Transactions of the American Philosophical Society, 1771-1808*, 6 vols. (New York: Kraus Reprint Corp., 1966) contain invention descriptions and drawings made by William Henry, John Sellers, Richard Wells, and others. The Simon Gratz Autograph Collection of manuscripts at the Historical Society of Pennsylvania also has some Fitch and Rumsey pieces and a smattering of items on other inventors. The financial records and manufacturing committee minutes for the Pennsylvania manufacturing society for 1787 to 1789 are also in the manuscript collection of the Historical Society of Pennsylvania. The rare book room of the Van Pelt Library at the University of Pennsylvania has a manuscript copy of John Beale Bordley's notes on communications to the Philadelphia agricultural society from 1785 to 1789.

The Library of Congress holds manuscripts touching on almost everything covered in this study. The Peter Force Papers are perhaps the single most important collection in the Library of Congress dealing with the Revolutionary Era. Force pulled together transcripts of manuscripts in British archives as well as original documents from committees of safety and provincial conventions in the colonies. Library of Congress miscellaneous manuscripts also include a wide variety of sources. In short, the Library of Congress is rich in resources, and its collections are nearly overwhelming.

I relied heavily on newspapers and magazines. The American Philosophical Society has an excellent run of newspapers, including the *Pennsylvania Gazette* and the *Pennsylvania Journal and the Weekly Advertiser*. The former has also been reprinted in twenty-five bound volumes, 1728-1799, by Microsurance Inc. (Philadelphia, 1968). I used the microfilm copy of the *Virginia Gazette* at the University of California, Santa Barbara, and microfilm copies of the *New York Journal* and *Boston Gazette and Country Journal* at the University of California, Berkeley. The American Philosophical Society and the Henry E. Huntington libraries have extensive collections of eighteenth-century American magazines, and I also perused the comprehensive collection published in the *American Periodical Series* (Ann Arbor: University Microfilms, 1942, 33 reels microfilm). The Huntington Library in particular has a fine collection of rare books and pamphlets; and the Charles Evans *American Bibliography* of literature, edited by Clifford Shipton and James E. Mooney, and reproduced on microcard as the *Early American Imprint Series* (Worcester: American Antiquarian Society and the Readex Microprint Corp., 1956-1968) is indispensable for research in early American history.

I also consulted numerous document collections for colonial and state governments. One of the most comprehensive series covers Virginia through the end of the Revolutionary Era. See William W. Hening, ed., *The Statutes at Large*, 13 vols. (New York: R. & W. Bartow, 1823, 2nd edition); H. R. McIlwaine and J. P. Kennedy, eds., *Journals of the House of Burgesses, 1619-1776*, 13 vols. (Richmond: Everett Waddey Co., 1905-1915); and H. R. McIlwaine et al., eds., *Journals of the Council of the State of Virginia, 1776-1788*, 4 vols. (Richmond: Division of Purchase and Printing, 1931-1967). Samuel Hazard, ed., *Pennsylvania Archives* (Harrisburg: Historical Society of Pennsylvania, 1852-); and idem, ed., *Pennsylvania Colonial Records* (Harrisburg: Historical Society of Pennsylvania, 1852) contain information on inventors recorded in the assembly and council of safety minutes for that state and nearly all of the surviving papers relating to the chevaux-de-frise.

For details on pre-war home manufactures and the munitions industry during the war, Peter Force, ed., *American Archives*, 9 vols. (Washington, D.C.: M. St. Clair-Clarke and Peter Force, 1837-1853) has hundreds of pertinent documents culled from newspapers and assembly proceedings from 1774 to 1776. Another invaluable documents series, specializing in commerce and naval warfare (and as yet unfinished) is *Naval Documents of the American Revolution*, edited by the Naval History Division, in 8 volumes to date, through 1778 (Washington, D.C.: Department of the Navy, 1964-). This collection draws from both European and American archives.

For other documents collected in Europe see Benjamin F. Stevens, ed., *Facsimiles of Manuscripts in European Archives Relating to American, 1773-1783*, 25 vols. (London: Chiswick Press, 1898), a large collection which represents only a fraction of the documents Stevens catalogued, almost all of which have been copied as transcripts deposited at the Library of Congress; Kenneth G. Davies, ed., *Documents of the American Revolution, 1770-1783*, 21 vols. (Shannon, Ireland: Irish University Press, 1972-); and Merrill Jensen, ed., *American Colonial Documents to 1776* (London: Eyre & Spottiswood, 1955). All three editors drew from the Public Record Office in London, and Stevens from other European depositories as well.

Julian P. Boyd et al., eds., *The Papers of Thomas Jefferson*, 20 vols. (Princeton: Princeton University Press, 1950-), now up to 1791, contains information on most of the individuals discussed in the later chapters of this book. For more specialized letter collections, journals, and diaries, the reader is referred to the notes at the end of each chapter.

Much has been done in the history of American technology over the past two decades. But as a survey of Brooke Hindle's recommendations for further study in his *Technology in Early America* (Chapel Hill: University of North Carolina Press, 1966) would show, there is still much to do. Hindle himself made an invaluable contribution with his encyclopedic *The Pursuit of Science in Revolutionary America, 1735-1789* (Chapel Hill: University of North Carolina Press, 1956). I agree with Hindle that, generally speaking, the war proved detrimental to the quick advance of American science and technology. I would add that if the war was harmful on one level it proved helpful on another by showing Americans just how far they had to go before they could even begin to rival European scientific and technological expertise.

Most early writing about colonial technology focused on individuals and the heroic inventor in particular. Henry Howe's biographical *Memoirs of the Most Eminent American Mechanics* (New York: Alexander V. Blake, 1841) is indicative of this approach; and Roger Burlingame's *March of the Iron Men* (New York: Charles Scribner's Sons, 1938), though written nearly a century later, is part of the same genre. Burlingame is nevertheless still worth reading. His book was a pioneering attempt to deal with invention in its social context. He erred in many particulars and created new myths even as he struck down others, but he did have a feel for the pervasive importance of technology in American society. When talking about the Revolutionary Era, John W. Oliver, *History of American Technology* (New York: Ronald Press, 1956), and I. Bernard Cohen, "Science and the Revolution," *Technology Review* 47 (1945):367-368, 374, 376, 378, showed their tie to Burlingame's style by presenting a few men and inventions as evidence of a more general technological genius. Even Dirk Struik, *Yankee Science in the Making* (Boston: Little, Brown and Co., 1948), and Carl Bridenbaugh's scholarly *The Colonial Craftsman* (Chicago: University of Chicago Press, 1950), are primarily catalogs recording the accomplishments of tinkerers and inventors rather than in-depth analyses of the social acceptance or rejection of technology. Joseph and Francis Gies' popular account *The Ingenious Yankees* (New York: Thomas Y. Crowell, 1976) is in some respects better.

One of the first articles to study the growth of technological awareness in the United States was Samuel Rezneck, "The Rise and Early Development of Industrial Consciousness in the United States, 1760-1830," *Journal of Economic and Business History* 4 (1932):784-811. A. Hunter Depree, *Science in the Federal Government* (Cambridge: Harvard University Press, 1957) looked at the growing commitment of the national government to science from the founding fathers through the end of World War II. Whitfield Bell, "Science and Humanity in Philadelphia, 1775-1790," (Ph.D. dissertation, University of Pennsylvania, 1947), and more especially Hugo A. Meier, "The Technological Concept in American Social History, 1750-1860," (Ph.D. dissertation, University of Wisconsin, 1950), gave attention to the social acceptance of technology in Revolutionary America. The Meier dissertation presaged John F. Kasson's *Civilizing the Machine* (New York: Grossman/Viking, 1976), with its observations on American thinking about technology in a republican society during the late eighteenth century and throughout the nineteenth century. Charles L. Sanford, *The Quest for Paradise* (Urbana: University of Illinois Press, 1961), puts the American obsession with republican industry into the mainstream of a search for the New World regenerative experience. Leo Marx, *The Machine in the Garden* (New York: Oxford University Press, 1964), is also valuable, though it does not deal as explicitly with American technology in the Revolutionary Era as do Meier and Kasson. Carroll W. Pursell, Jr., *Early Stationary Steam Engines in America* (Washington, D.C.: Smithsonian Institution Press, 1969), is a pioneer of its type, exploring the status of steam technology in the antebellum United States. Silvio A. Bedini, *Thinkers and Tinkers* (New York: Charles Scribner's Sons, 1975), is a useful study of scientific instruments and instrument makers in the early republic. It offers a number of important insights but is stylistically like the studies of Bridenbaugh and Struik. Brooke Hindle's *Emulation and Invention* (New York: New York University Press, 1981) discusses most imaginatively the meaning of invention to Americans of the early national period.

One recent and much-needed study questioning some of the time-honored theories regarding American receptivity to innovation is Merritt Roe Smith,

Harpers Ferry Armory and the New Technology (Ithaca: Cornell University Press, 1977), a sophisticated appraisal of one "industrial" community's reaction to technological change. Also see James A. Henretta, "Families and Farms: *Mentalité* in Pre-Industrial America," *William and Mary Quarterly*, 3rd series, 35 (1978):3-32; and Gary B. Kulik, "The Beginnings of the Industrial Revolution in America: Pawtucket, Rhode Island, 1672-1829" (Ph.D. dissertation, Brown University, 1980), the latter of which shows that industrial transformation was gradual and, if the new textiles technology helped reshape society, that technology itself had to fit into a larger social system. Hence the danger of talking too loosely about "modernization" or "transformation," in Pawtucket or anywhere else. Kulik, Smith, and Anthony F. C. Wallace, *Rockdale* (New York: Alfred A. Knopf, 1978) have all shown the importance of local histories to our understanding of more amorphous national developments.

Trying to trace the source of an individual's genius is perhaps impossible, but we need to study more closely the social reaction to invention to see how and why Americans began to identify with invention and technology in general. Carroll W. Pursell, Jr., "Thomas Digges and William Pearce: An Example of the Transit of Technology," *William and Mary Quarterly*, 3rd series, 21 (1964):551-560; and Norman B. Wilkinson, "Brandywine Borrowings From European Technology," *Technology and Culture* 4 (1963):1-13, pointed the way by showing the importance of technological transfer to the early republic.

More recently David J. Jeremy studied the flow of technology to the United States in a series of articles: "Innovation in American Textile Technology during the Early 19th Century," *Technology and Culture* 14 (1973):40-76; "British Textile Technology Transmission to the United States: The Philadelphia Region Experience, 1770-1820," *Business History Review* 47 (1973):24-52; "Damming the Flood: British Government Efforts to Check the Outflow of Technicians and Machinery, 1780-1843," *Business History Review* 51 (1977):1-34; and with Anthony F. C. Wallace, "William Pollard and the Arkwright Patents," *William and Mary Quarterly*, 3rd series, 34 (1977):404-425. Jeremy has brought together many of his conclusions in *Transatlantic Industrial Revolution* (Cambridge: MIT Press, 1981). The studies of Pursell, Wilkinson, Jeremy, and Paul E. Rivard, "Textile Experiments in Rhode Island, 1788-1789," *Rhode Island History* 33 (1974):35-45, offer ample evidence of the complexity of technological change in Revolutionary America.

Useful studies on the Industrial Revolution in Great Britain are T. S. Ashton, *The Industrial Revolution, 1760-1830* (London: Oxford University Press, 1948); Paul Mantoux, *The Industrial Revolution in the Eighteenth Century* (1928; reprint ed., New York: Harper and Row, 1961) Phyllis Deane, *The First Industrial Revolution* (Cambridge: Cambridge University Press, 1965); A. E. Musson and Eric Robinson, *Science and Technology in the Industrial Revolution* (Toronto: University of Toronto Press, 1969); Peter Lane, *The Industrial Revolution* (New York: Barnes and Noble, 1978); and David S. Landes, *The Unbound Prometheus* (Cambridge: Cambridge University Press, 1969). For American industry during the Revolutionary Era, see the general accounts in James L. Bishop, *A History of American Manufactures, 1607-1860*, 2 vols. (Philadelphia: Edward Young and Co., 1864); Victor S. Clark, *History of Manufactures in the United States, 1607-1928*, 3 vols. (New York: Carnegie Institute of Washington, 1929); Rola M. Tryon, *Household Manufactures in the United States, 1640-1860* (Chicago: University of Chicago Press,

1918); William R. Bagnall, *The Textile Industries of the United States*, 2 vols. (Cambridge: Riverside Press, 1893); and the briefer overview in Elisha P. Douglass, *The Coming of Age of American Business* (Chapel Hill: University of North Carolina Press, 1971).

These general studies, the works on the Industrial Revolution in Britain, and more specialized accounts like Robert A. East, *Business Enterprise in the American Revolutionary Era* (New York: Columbia University Press, 1938); Thomas C. Cochran, *Frontiers of Change* (New York: Oxford University Press, 1981); a masterful gathering of writings on the American system of manufactures in Otto Mayr and Robert C. Post, *Yankee Enterprise* (Washington, D.C.: Smithsonian Institution Press, 1981); and David A. Hounshell, *From the American System to Mass Production* (Baltimore: Johns Hopkins University Press, 1984), can serve as starting points for more detailed looks at how technology grabbed hold in the United States. These studies need to be read along with others that deal with the political, social, and intellectual life of early America, studies that do not necessarily deal directly with technology but that nevertheless help to put it into the proper context. Thus the value of J. E. Crowley, *This Sheba, Self* (Baltimore: Johns Hopkins University Press, 1974); Drew R. McCoy, *The Elusive Republic* (Chapel Hill: University of North Carolina Press, 1980); and Ralph Lerner, "Commerce and Character: The Anglo-American as New Model-Man," *William and Mary Quarterly*, 3rd series, 36 (1979):3-26, all of which reflect a renewed interest in political economy. As these studies show, the history of technology is part of a much larger whole. If it suffers when ignored, it also suffers when studied in isolation.

INDEX

Adams, John: on munitions, 64, 72, 77; wartime manufactures, 68; weakness of Congress, 88; with John Macpherson, 89-90, 91; and David Bushnell, 95; Delaware River chevaux-de-frise, 121

Almy, Brown and Slater, textiles manufacture, 167-68, 172

Amelung, John, glass manufacturer, 173-74

American Academy of Arts and Sciences, 184-85

American Philosophical Society: promotes science and technology, 41-43; silk growing and canals, 43; limited promotion of invention, 49; nationalism of, 55-56; post-war activities, 184, 185-86; Fitch-Rumsey dispute, 196, 198-99

"American Turtle," 92-98, 100, 107, 214. *See also* Bushnell, David

Anderson, Ephraim, fireships, 105-6, 120

André, John, Delaware River chevaux-de-frise, 126

Arkwright, Richard, water frame, 44, 45, 157, 158, 166, 167–68

Arnold, Benedict, and Joseph Belton, 102

Articles of Confederation, patents under, 192, 193-94, 198

Augusta, British warship sunk in Delaware River, 126

Barlow, Joel, promise of new nation, 205

Barnard, Thomas, manufactures and imported technology, 17-18, 28, 40

Barnes, Joseph: in Fitch-Rumsey dispute, 199; complaints about federal patent law, 202-3

Barton, William, post-war manufactures, 156

Bauman, Sebastian, gunnery experiment, 162

Beaufort, Comte de, manufacturing proposal, 162

Beaumarchais, Pierre-Augustin Caron de, aid to patriots, 76-77, 79

Begozzat, Pierre, wartime trade with patriots, 75

Belton, Joseph: submarine of, 100-101, 120; rapid-fire musket, 101-2, 104, 107

Bemis Heights, battle of, riflemen at, 143

Bernard, Governor Francis, colonial manufactures, 26

Beverly, Mass., textiles experiment, 166

Biddle, Owen: in American Philosophical Society, 41; silk growing, 43; and Arthur Donaldson, 49; Delaware River defenses, 118; post-war science, 184

Bigelow, Jacob, nation of inventors, 222

Billingsport, fort at, 122, 123-24, 126
Bliss, John, Leffingwell bridge, 187
Board of Trade, uneven restriction of technology export: pre-war, 16, 39, 44; post-war, 160, 218
Board of War (and Ordnance): munitions production, 68, 77; shortcomings, 71; disillusioned with rifles, 141
Bordley, John Beale: promotes agriculture, 159-60, 162; manufactures, 163, 170
Bond, Phineas, on American ambitions, 172-73
Bond, Dr. Thomas: pre-war manufactures, 20; American Philosophical Society, 41; post-war science, 184
Boorstin, Daniel, and "givenness," 6
Boston Gazette, notes riflemen, 137
Boulton, Matthew: steam technology, 44, 45; and James Rumsey, 199
Bowdoin, Governor James, post-war science, 185
Bowler, Metcalf, improved farming, 158, 160
Bridenbaugh, Carl, on colonial craftsmen, 38
Briggs, Samuel, inventor, 194
Brown, Moses, entrepreneur, 167-68, 172, 175, 220
Brown, Richard D., on modernization, 216-17
Bull, John, design of Fort Mercer (Red Bank), 120, 123
Bull, William, predicts soil depletion, 11
Burgoyne, General John, on riflemen, 143
Burlingame, Roger: on Yankee ingenuity, 100; rifle in war, 132; ideas critiqued, 219-20
Bushnell, David: submarine design, 91-96, 101; attempt on HMS *Eagle,* 97-98; explosive mines, 98-99; disappointments of, 99-100; significance of, 107, 214-15
Butler, Thomas, armorer, 68
Cabot, George, Beverly textiles experiment, 166
Capellen, Baron Van Der, and Stephen Sayre, 103
Carey, Matthew: *American Museum* and post-war manufactures, 156; invention, 221-22
Carpenters Company, 116, 117
Carroll, Daniel, supports John Amelung, 173-74
Centennial Exposition, celebration of American technology, 222-23
Cerberus, attacked by Bushnell, 98
Chastellux, Francois Jean, Marquis de: Delaware River defenses, 125; promise of new nation, 205-6
Chauncey, Elihu, flax dressing patent, 47-48
Chevaux-de-frise. *See* Delaware River defense system
Christiansen, Hans, water supply system, 48
Churchman, John, inventor, 194
Cochran, Thomas C., colonial American technology, 29
Cohen, I. Bernard: wartime problems, 63; government and science, 106
Coke, Sir Edward, on patents, 200
Coleman, Robert, ironmaster, 121-22
Colles, Christopher, mechanical genius: before war, 51-56; during war, 87; after war, 193-94, 198; and John Fitch, 197
Committee of Secret Correspondence, imports munitions, 74
Constitution: promotes manufactures, 165, 172, 173-74; invention, 194-95
Continental army. *See* Pennsylvania rifle
Continental Association, and home manufactures, 21-22
Continental Congress: home manufactures, 21-22; accepts Penet-Coulaux manufacturing offer, 78-79; and wartime invention, 88-89, 104, 106-7; recruits engineers, 122; Dela-

ware River defenses, 123; raises riflemen, 136-37, 138, 148; rejects Beaufort manufacturing offer, 162; and post-war invention, 192-94. *See also* Articles of Confederation; Board of War; Committee of Secret Correspondence; Munitions production

Cort, Henry, iron innovation, 158

Coudray, Phillipe Charles, Tronson du, engineer, 123-24

Cox, Lemuel, bridge design, 187

Coxe, Tench, promotion of: manufactures, 156, 157-58, 162, 164, 166, 200; agriculture, 158, 160, 170; invention, 186; federal patent law, 201; prophesies manufacturing success, 175; *See also* Pennsylvania Society for the Encouragement of Manufactures; Society for the Encouragement of Useful Manufactures

Crèvecoeur, Hector St. John de, plight of inventors, 200, 203-4

Crompton, Samuel, mule inventor, 158, 165

Cruse, Englehart, inventor, 194

Davis, Thomas, carpenter, 122

Deane, Silas, congressional agent, 74; Bushnell and, 95; post-war textiles, 158

Declaration of Independence, 56, 68, 221

Delaware River defense system: plans for, 113-22; chevaux-de-frise, 117-19, 121, 122-24; pilots for, 118; firerafts and navy, 120; forts along, 120; Belton submarine rejected, 120; French engineers, 122-24; campaign against, 124-25; collapse, 125-27

Dewey, Joseph, steel patent, 47

Dickinson, John, maligned by Macpherson, 89; and John Fitch, 210 n.60; need for invention, 213-14

Digby, William, on riflemen, 143

Donaldson, Arthur, inventor of dredge, 49; Delaware River defenses, 116, 122, 123; Hudson River defense, 122; post-war dredge patent, 185, 211 n.81; Fitch-Rumsey dispute, 197

Drowne, Solomon, on Delaware River defenses, 121-22

Dulany, Daniel: manufactures to ease imperial tension, 16; problem with that view, 16, 218

Dwight, Timothy, eulogizes Bushnell, 92

Eagle, attacked by "Turtle," 97

East, Robert, on wartime business activity, 63

Edison, Thomas, inventor as folk hero, 206, 219

Eliot, Jared, scientific farming, 9, 10; black sand essay as invention, 46

Ellicott, Andrew, entrepreneur, 175

Ellicott family, innovative millers, 10, 189, 208 n.25

Engineers. *See* Delaware River defense system; Smith, Robert

Evans, Oliver: wartime invention, 88; flour milling machinery, 185, 187-91; steam carriage, 185, 190; federal patent, 202; criticizes patent law, 202, 203; problems not caused by, 204; question of originality, 208 n.25; appeal of milling innovations, 215-16, 220; invention as a science, 218-19; venerated, 221

Farming, techniques slow to change, 9-10, 11-12, 28, 30 n.5, 159-60, 163, 168-71

Fauquier, Governor Francis, on colonial manufacturing, 26

Fenton, Jotham and Nathan, inventors, 186

Ferguson, Eugene, 3; on Yankee ingenuity, 220

Ferguson, Major Patrick, breechloading rifle, 153-54 n.60

Fisher, Joshua, hydrographic survey, 115
Fitch, John: steamboat dispute with Rumsey, 189, 195-201; criticizes federal patent law, 202; pride in role of inventors, 204-5, 221
Fleury, Francois Louis de, French engineer, 124, 125, 127
Flowers, Benjamin, artificer regiment, 68-69
Foreign aid. *See* munitions production
Franklin, Benjamin: urges manufacturing, 16; notes lack of progress, 26; as inventor, 38; misgivings about manufactures, 40, 58 n.12, 176 n.6; and Royal Society for the Encouragement of the Arts, 41; American Philosophical Society, 41; promotion of invention, 42; lightning rod as important invention, 46; complains inventors ignored, 55; munitions shortages, 72; on Bushnell's "Turtle," 95; Belton's submarine, 100, 101; Delaware defenses, 117, 118, 126; dislikes muskets, 148; and English textiles workers, 160-61; advice to immigrants, 161; caution to enthusiasts for change, 171; Fitch-Rumsey dispute, 197-98
Fraser, General Simon, slain by riflemen, 143
Freeman's Farm, battle of, riflemen at, 143
Frontier, impact on rate of technological change, 17-18, 40, 156
Fulton, Robert, inventor as hero, 206

Gage, General Thomas: Pennsylvanians manufacturing, 20; rebels importing munitions, 73
Gale, Dr. Benjamin, and Bushnell's "Turtle," 95
Galvez, Don Bernardo de, munitions for Americans, 75
Gates, Horatio, and Joseph Belton, 102
Germantown, Pa., manufacturing society, 165-66, 170
Godfrey, Thomas, inventor, 46
Govett, Joseph, carpenter, 117
Great Bridge, battle of, riflemen at, 139
Greene, Nathanael: Delaware River defenses, 113, 124; at Guilford Courthouse, 146
Greer, Arthur, inventor, 194
Guilford Courthouse, battle of, riflemen at, 146
Gunpowder. *See* munitions production
Gunsmiths. *See* munitions production

Habakkuk, H. J., on Yankee ingenuity, 219
Hague, John, inventor, 48-49, 56
Halley, Sir Edmund, theory on underwater explosives, 92
Hamilton, Alexander, need for manufactures, 155, 158, 160-61, 168, 174, 200, 204. *See also* Society for the Encouragement of Useful Manufactures
Hamlin, Jabez, flax dressing patent, 47-48
Hammond, George, on American manufacturing, 171-72
Hancock, John, marksmanship of riflemen, 136
Hannah's Cowpens, battle of, riflemen at, 145-46
Harbaugh, Leonard, inventor, 194
Hargreaves, James, spinning jenny, 45, 48, 168
Harrison, Benjamin, shortage of engineers, 113
Hasenclever, Peter, ironmaster, 13-14
Heaton, Herbert, on post-war manufactures, 165
Henretta, James, colonial economic self-sufficiency, 27
Henry, William, forgotten gunsmith, 136
Higley, Samuel, steel patent, 47

Hillsborough, Wills Hill, 1st Earl of, colonial secretary, 64
Hindle, Brooke, 3; wartime science, 87, 106; post-war steamboat technology, 201
Hitchcock, Gad, sermon on military arts, 142
Hobday, John, thresher inventor, 49-50, 55, 56; during war, 87
Hodge, William, congressional agent, 74
Home manufactures movement: threat to empire, 12-13, 16-17; beginnings, 17-24; limitations on, 24-26; not complete failure, 26-27; growing aspirations, 28-29, 56; post-war movement, 155-58; limits on, 164-68; promise of future, 172-76
Hopkins, James, inventor, 193, 195
Hopkinson, Francis: science in colonies, 42; "Battle of the Kegs" and Bushnell, 99; post-war science, 184; criticizes inventors, 192; Fitch-Rumsey dispute, 197
Howe, Lord Richard: flagship *Eagle* attacked by "Turtle," 97; Delaware River campaign, 124-25
Howe, General William: rebels receiving contraband, 75; defeats Washington in Pennsylvania, 124
Hunter, James, iron foundry, 67

Industrial Revolution: and increased identification with invention, 6, 38, 218; colonists largely unaware of, 37-39, 44-45; patents and, 47; difficulty generalizing about, 58 n.11, 106, 177 n.9; growing American awareness, 156-57, 214, 218
Invention: not promoted vigorously in colonies, 6, 37-38, 45-47, 53-57; source of national pride, 6, 221-23; perceived as curiosity, 44-47; wartime limits on, 87-89, 106-7; after war, 214, 218-20. *See also* Colles, Christopher; Evans, Oliver; Yankee ingenuity
Irish, Nathan, and Delaware River defenses, 120
Iron Act of 1750, 14, 17; violations of, 21
Iron Industry: bone of imperial contention, 13-14; natural limits on, 17; post-war, 168

jaegers, German riflemen, 140
Jay, Sir James: gunnery plan, 106; post-war manufactures, 156
Jefferson, Thomas: war distracts scientists, 87; federal patent law, 200, 202; technological ambivalence, 200-201, 217-18
Jenkes, Joseph, early patent, 47
Johnson, Samuel, 8
Johnson, Governor Thomas, Fitch-Rumsey dispute, 196-97
Jones, Robert, oration on manufactures, 81
Joy, Daniel, on Delaware River defenses, 121-22

Kasson, John, *Civilizing the Machine*, 7 n.4
King's Mountain, battle of, riflemen at, 145
Knox, Henry: munitions production, 68; and Macpherson, 90, 192; and Stephen Sayre, 104-5; on Morgan's riflemen, 143
Kosciuszko, Thaddeus, and Delaware River defenses, 122-23

Lee, Arthur, and Stephen Sayre, 103
Lee, Charles: dislikes riflemen, 138; respects rifles, 139; plan for training American troops, 142
Lee, Richard Henry, 49; need for foreign aid, 74; superiority of riflemen, 136
Leslie, Robert, inventor, 178-79 n.36
Lemmon, Robert, inventor, 185
Lemon, James T., on colonial farming, 30 n.5
Lewis, Andrew, inventor, 194
Lewis, John, and Bushnell, 95

Livingston, Chancellor Robert, post-war agriculture, 171
Livingston, Governor William: munitions shortages, 72; and Stephen Sayre, 103
Logan, George, ambivalent support of manufactures, 167, 170, 217
Long Island, battle of, riflemen at, 140

McCoy, Drew R., *The Elusive Republic,* differing view of receptivity to manufactures, 58 n.12, 176-77 n.6
McMechen, James, and steamboat dispute, 195
Macpherson, John, inventor: erratic behavior during war, 89-91; compared with Bushnell, 91-92; Delaware River defense, 120; post-war inventions, 192, 194
Madison, James, patent clause in Constitution, 194
Magellan, John de Hyacinth, reward to inventors, 186
Manufactures. *See* Home manufactures movement
Manufacturing Society of Williamsburg, short-lived, 71
Marion, Francis, and riflemen, 145
Marshall, Thomas, textiles machinery, 167
Marx, Leo, "middle landscape," 218
Mascarene, John, manufactures to ease imperial stress, 15-16
Massachusetts Society for Promoting Agriculture, 171
Massie, William, thresher inventor, 50
Matlack, Timothy, on post-war science, 184
Mauduit du Plessis, Thomas-Antoine, French engineer, 124, 127
Maxcy, Jonathan, inventors as nation builders, 221
Meier, Hugo, 3; wartime industry, 63
Merlin, British ship sunk in Delaware River, 126
Miller, Aaron, inventor, 49
Mitchell, Andrew, industrial espionage, 158
Moore, Governor Henry, lack of colonial manufacturing, 26
Moore's Creek Bridge, battle of, riflemen at, 139
Morgan, Daniel: fame of, 136, 137; special rifle regiment under, 139-40, 142-44; at Hannah's Cowpens, 145-46. *See also* Pennsylvania rifle
Morris, Richard B., on wartime labor, 71-72
Moultrie, General William: munitions shortage, 65; drawbacks of Fabian tactics, 147
Mud Island (Fort Mifflin), 120-27
Muhlenberg, Colonel Peter, dislikes rifles, 141
Mumford, Lewis, 7; Americans and "megamachine," 224 n.10
Mumford, Thomas, wartime trade, 75
Munitions production: early inadequacy of, 64-65; organized by states, 65-67; by Congress, 67-69; how to increase, 69; lack of standardization and inefficiency, 69-72; shortages, 72, 77-79; imported from Europe, 72-77; totals, 85 n.69. *See also* Penet, P.: and Coulaux, J.
Murray, Sir James, on Delaware River defenses, 125
Muskets, *See* Munitions production; Pennsylvania rifle

New York Society for the Promotion of Agriculture, Arts, and Manufactures, 166
New York Society for the Promotion of Arts, Agriculture, and Economy, 20, 23, 40-41
Nicola, Lewis, Delaware River defense plan, 114-15
North, Douglass C., on economic growth, 4

Page, John, need for wartime scientific activity, 88

Paine, Robert Treat, bad gunpowder, 70
Paine, Thomas: and American Philosophical Society, 43; promotes invention, 55-56; wartime munitions production, 69, 77; steel crossbow, 88; argues war advanced science, 106; bridge design, 187, 200
Parkinson, George, textiles machinery, 167, 202
Parsons, Samuel, and Bushnell, 97
Patents: in the colonies, 47-48; in states, 185, 192; under Articles of Confederation, 192, 193-94; federal, 194-95, 201-4; Fitch-Rumsey dispute, 195-200. *See also* Invention; Evans, Oliver; Yankee ingenuity
Peale, Charles Willson: and Joseph Belton, 102; rifling experiment, 147
Pearce, William, textiles machinery, 167
Penet, P.: early trade with patriots, 75; and Coulaux, J., scheme to manufacture munitions, 78-79, 80
Pennsylvania committee of safety, Delaware River defense, 114, 115, 117, 118
Pennsylvania Gazette: vague notion of invention before war, 46-47, 136; clearer ideas after war, 157
Pennsylvania General Assembly: aid to Rittenhouse, 42; limited pre-war aid to inventors, 49, 51; Delaware River defense, 114; promotes postwar textiles, 164; Oliver Evans and, 189; Fitch-Rumsey dispute, 197
Pennsylvania rifle: legend of, 132; cost of and difficulty of making, 133, 148; evolution of, 133-34; compared with musket, 134-36; Pennsylvania, Virginia and Maryland rifle regiments, 136-38, 139; feared, 139; not feared, 140; disliked, 141; phased out of Continental army, 141-42; as symbol of Yankee ingenuity, 148; turned to in desperation, 214
Pennsylvania Society for the Encouragement of Manufactures and the Useful Arts, rise and fall, 164-65
Philadelphia Society for Promoting Agriculture, 170
Pickens, Andrew, and riflemen, 145
Pinckney, Charles, on patents, 194
Pliarne, Emanuel de, aid to patriots, 75
Poellnitz, Baron, scientific farmer, 160
Pollard, William, textiles machinery, 166, 202
Pollock, Oliver, congressional agent, 74
Potter, David, on abundance, 7 n.4
Potts, Stacy, inventor, 193, 195
Powel, Samuel, and manufactures, 170
Pursell, Carroll W., Jr., steam technology in colonies, 45

Quebec, American failure at, 139
Quincy, Josiah, on galleys, 90

Ramsay, David, science in new nation, 183-84
Raynal, Abbé, writings, 160
Red Bank (Fort Mercer), 120-27
Reed, Joseph, need to defend Delaware River, 114
Richardson, Enoch, spinning jenny made by, 161
Riflemen. *See* Morgan, Daniel; Pennsylvania rifle
Rittenhouse, David: inventor, 38; American Philosophical Society, 41; and Arthur Donaldson, 49; Christopher Colles and, 53; Joseph Belton and, 102; rifling experiments, 147; during the war, 160; criticizes inventors, 192-93; Fitch-Rumsey dispute, 197
Robinson, Ebenezer, Delaware River defense plan, 115-16
Roderique Hortalez et Cie., aid to patriots, 76-77

Rosenberg, Nathan, technological efficiency relative, 30 n.5
Rostow, Walt, economic takeoff, 175
Royal Society for the Encouragement of Arts, Manufactures and Commerce, 23; pattern for American Philosophical Society, 41, 49, 56; ties to, 42, 60 n.33
Rumsey, James, steamboat experiments: dispute with Fitch, 189, 195-200; criticizes federal patent law, 202
Rush, Dr. Benjamin: pre-war manufactures, 22-23; fear of large-scale manufactures, 40; wartime munitions production, 69; post-war manufactures, 156-57, 170, 173

Sawmills: early mechanization in colonies, 8; no threat to empire, 11; in response to market, 26; innovative, 29, 38
Sayre, Stephen, ship and gunnery innovation, 102-5, 107
Schlesinger, Arthur M.: limited objectives of manufacturing enthusiasts, 24; conclusion questioned, 24-25
Schultz, William, inventor, 186
Sellers, John, inventor, 51
Sewall, Samuel, bridge builder, 187
Sharrer, George Terry, on automated milling, 208 n.25
Sheffield, John Baker Holroyd, 1st Lord, predicts American manufacturing mediocrity, 175
Shelburne, William Petty, 2nd Earl of, colonial secretary, 20
Shipman, John, flour mill patent, 48
Simcoe, John, riflemen not fearsome, 147
Siracusa, Carl, on technological consciousness, 7 n.3
Slater, Samuel, symbol of Yankee ingenuity, 220, 222. *See also* Almy, Brown and Slater
Smith, Adam, on American desires to manufacture, 32 n.21, 32-33 n.29
Smith, Merritt Roe, on technological ambivalence, 223 n.5
Smith, Robert, architect and engineer, 116; chevaux-de-frise for Delaware River, 117-20; death of, 122; importance, 122, 123, 127, 215. *See also* Delaware River defense system
Smith, William: home manufactures, 20; role of American Philosophical Society, 42
Society for Encouraging Industry and Employing the Poor, 17
Society for the Encouragement of Useful Manufactures (SUM), 166-67
Somers, Thomas, textiles manufacture, 166
Stanly, John, munitions import estimate, 76
Steam engines, in colonies, 45, 53-55
Stephens, Thomas, potash making, 21
Steuben, Baron Friedrich Wilhelm von, rebuilds American army, 141-42
Stevens, John, wartime gunnery scheme, 105
Stewart, James, technological transfer, 41
Stiegel, Henry, glass factory, 25
Stiles, Ezra: home manufactures, 18; Bushnell's "Turtle," 95; post-war invention and science, 186; promise of national future, 205-6
Stoebel, John, inventor, 194
Stormont, David Murray, 7th Viscount, patriots importing munitions, 78
Submarines, French model, 45. *See also* Belton, Joseph; Bushnell, David
Sumter, Thomas, and riflemen, 145

Tarleton, Banastre, 145
Technological change: American identification with, 3-7, 205-6, 213-23; rate of, 6, 182 n.74, 215-17; desire for, 26-29, 37; difficulty of during war, 63-64, 80-81; rate of

change less significant than desire for, 175-76; ambivalent attitude toward, 176-77 n.6, 217, 223 n.5
Textiles: in colonies, 22-24, 49, 56-57; after the war, 80-81, 157-58, 160-61, 164-68. *See also* Coxe, Tench; Pennsylvania Society for the Encouragement of Manufactures; Slater, Samuel; United Company of Philadelphia for Promoting American Manufactures
Thompson, Benjamin (Count Rumford), disdain for riflemen, 138
Thompson, Colonel William, commander of Pennsylvania riflemen, 137
Timmons, John, patent for rice cleaner, 48
Towers, Robert, Pennsylvania commissary, 67
Treaty of Amity and Commerce, 77
Trumbull, Governor Jonathan, and Bushnell's "Turtle," 96, 99
Tully, Christopher, spinning jenny, 48-49, 56
Turner, George, inventor, 186
"Turtle." *See* "American Turtle"

United Company of Philadelphia for Promoting American Manufactures, 22-24, 48; collapse, 80; hope for future, 81, 164. *See also* Pennsylvania Society for the Encouragement of Manufactures

Van Doren, Carl, on Franklin and chevaux-de-frise, 117
Varlo, Charles, English agronomist, 160
Vergennes, Charles Gravier, Comte de, aid to Americans, 76
Vigilant, British warship, 125, 126
Virginia Society for Advancing Useful Knowledge: organized, 43, 50, 55; collapses, 87-88
Voight, Henry, craftsman, 197

Wall, George, patent to, 185
Walker, George, thresher inventor, 50
Wansey, Henry, state of American manufactures, 167
Washington, George: pre-war farming experiments, 9-10, 28; munitions shortages, 72, 78; John Macpherson and, 90; on David Bushnell, 97; Joseph Belton and, 101; Stephen Sayre, 103, 104; John Stevens, 105; Ephraim Anderson, 105; James Jay, 106; engineer shortage, 113; Delaware River defenses, 123-24; attachment to riflemen, 136, 137-38, 140, 142-43, 146; limits to attachment, 138, 141, 146; post-war farming and manufactures, 162-63, 171; Evans mill machinery, 190; Christopher Colles canal plan, 193; Fitch-Rumsey dispute, 195, 196; federal patent law, 201
Watkeys, Henry, gunsmith, 67
Watt, James: steam engine, 44, 45, 57, 157; and James Rumsey, 199
Wayne, General Anthony, dislikes rifles, 141
Webster, Noah: federal patent law, 201; ambivalence toward manufactures, 217
Weeden, William, colonial quest for economic independence, 16
Wells, Richard: inventor, 51, 55; Christopher Colles and, 53; United Company of Philadelphia, 62 n.60; Delaware River defense plan, 115; post-war manufactures, 170; invention, 186; Fitch-Rumsey dispute, 197; criticizes federal patent law, 202
Wentworth, Governor John, pre-war manufactures, 26
White, Elisha and Robert, import skilled workers, 40
White, Lynn, 7; society and technological change, 7 n.1
Whitney, Eli: arms manufacture, 174, 219; cotton gin, 217, 220
Whyte, Robert, merchant, 118

Wilkinson, David, craftsman, 167
Wilkinson, Oziel, inventor, 88, 167
Wily, John, home manufactures, 21
Winthrop, John: theory on comets, 46; gunpowder making, 69
Wisner, Henry, gunpowder manufacture, 67
Wyld, Henry, English textile worker, 160-61
Yankee ingenuity: pride in, 3-7; in colonies, 37-38, 55-56; after the war, 183, 219-20. *See also* Invention; Patents; Technological change
Young, Arthur: treatise on farming, 9, 10, 160; respect for George Washington as scientific farmer, 163

About the Author

Neil Longley York is Associate Professor of History and Coordinator of the American Studies Program at Brigham Young University. His articles have appeared in *Military Affairs, Pennsylvania Magazine of History and Biography, National Forum,* and the *Journal of American Culture.*